NUCLEAR STRATEGY
India's March Towards Credible Deterrence

Dr Manpreet Sethi

Introduction by

Air Commodore **Jasjit Singh** AVSM VrC VM (Retd)

KW Publishers Pvt Ltd
New Delhi

in association with

Centre for Air Power Studies
New Delhi

Reprint, August 2025

The **Centre for Air Power Studies** is an independent, non-profit, academic research institution established in 2002 under a registered Trust to undertake and promote policy-related research, study and discussion on the trends and developments in defence and military issues, especially air power and the aerospace arena, for civil and military purposes. Its publications seek to expand and deepen the understanding of defence, military power, air power and aerospace issues without necessarily reflecting the views of any institution or individuals except those of the authors.

Jasjit Singh
Director
Centre for Air Power Studies
P-284, Arjan Path
Subroto Park
New Delhi 110010

Tele: (91-11) 25699131
E-mail: office@aerospaceindia.org

Published in India by
Kalpana Shukla
KW Publishers Pvt Ltd
NEW DELHI: 4676/21, First Floor, Ansari Road, Daryaganj, New Delhi 110002
T:+91.11.43528107 E:kw@kwpub.in W:www.kwpub.in

ISBN 10: 81-87966- 70-X
ISBN 13: 978-81-87966-70-8

Printed and Bound in India
Quiinkit Infotech Pvt. Ltd., Gurugram

To

Keerat, my son

So that his generation
may live without fear of nuclear holocaust

Contents

Preface

May 1998 was a momentous event in India's life. At one level, the five nuclear tests marked the culmination of the long debate on India's nuclear status. At another level, they initiated the country's journey towards credible nuclear deterrence. In the eleventh year of its existence as a state with nuclear weapons, India is engaged in a range of activities to meaningfully integrate the nuclear weapon into its national security strategy. The US and the USSR, at the same stage of their lives were engaged in pretty much a similar exercise. However, unlike the superpowers, for whom the immense destructive potential of the atomic weapon and its implications for inter-state relations suddenly burst on the scene and sent them scrambling to craft strategies that could fit the new reality, India's acquisition of nuclear weapons capability, though dictated by circumstances, came with a basic understanding of the ground rules of the game of nuclear deterrence. In fact, it may be recalled that soon after the tests, there was a spate of books and studies on India's nuclear strategy. A draft nuclear doctrine was made public just fifteen months after the tests and operational issues were beginning to be grappled with.

Much ground has since been covered. In the preface of his book written in 1993, Gen K. Sundarji had lamented, "The really big secret is that India has no coherent nuclear weapon policy and worse still, she does not even have an institutionalized system for analyzing and throwing up policy options in this regard."[1] The situation has indeed changed since then. Nevertheless, a lot more still needs to be done and establishing credible nuclear deterrence is work in progress. In fact, one can well argue whether it can ever be possible to achieve a state of 'perfect' credible nuclear deterrence, but a nation can, and must, certainly aspire for it by optimally developing the building blocks that are required. Given that the security of the nation and its citizens is the primary duty of the state, it must undertake this exercise because nuclear deterrence, pending universal elimination of nuclear weapons, is the only way of ensuring security in the face of an adversary's nuclear weapons. The stark reality is that there is no other defence against these weapons and the only way to avoid their being used against oneself is the assured promise of retaliation with the same – however unthinkable and distasteful that thought might be.

Credible nuclear deterrence demands the availability and meaningful integration of a few elements that collectively constitute the nuclear arsenal. The mere acquisition or demonstration of a nuclear weapons capability through a test does not make for credible deterrence. At the most, it provides existential deterrence by injecting a seed of uncertainty in the mind of the adversary and forcing him to grant the benefit of the doubt to the demonstrator country. It is akin to making a public announcement of capability, but it is for the adversary to judge whether that capability is convincing enough to cause him damage of the kind that he would consider unacceptable and, hence, be deterred. Credible nuclear deterrence, on the other hand, rests on the removal of all doubt from the mind of the adversary that any misadventure by him *would result in assured retaliation* of the kind that would inflict unacceptable costs on him.

1. Gen K. Sundarji, *Blind Men of Hindoostan: Indo-Pak Nuclear War* (New Delhi: UBSPD, 1993), p.xv.

Three basic elements are critical to constitute credible nuclear deterrence. To explain it with a simple analogy, credible nuclear deterrence may be visualised as a three-legged stool in which each leg symbolises *capability, resolve to use that capability,* and the *communication of both* — the capability and the resolve — to the adversary. Each of the three props is equally important and if they are in harmony, the stool, or nuclear deterrence, would be secure and credible. The requirement then boils down to adequately buttressing the three components of deterrence through measures such as:

- conceiving a clear role for the nuclear weapon and delineation of circumstances of its employment;
- evolving a targeting plan and building the necessary stockpile of warheads;
- developing a reliable capability to make the warhead deliverable on pre-determined targets;
- taking steps to ensure the survival of all elements of the arsenal in order to inflict unacceptable damage on the enemy;
- instituting adequate command, control and communication systems to ensure that the weapon is launched when authorised, and only when authorised.

The presence of all the above should be expected to deter a rational and reasonable adversary from initiating action that runs the risk of bringing nuclear weapons into use.

India has been engaged in the process of operationalising its nuclear deterrent over the last decade, with varying degrees of speed of activity, levels of transparency and measures of success. Some steps of the exercise such as the development, testing and induction of delivery vehicles (missiles) have been far more visible than others such as the development of plans, procedures and organisations necessary for the conduct of effective nuclear operations. India's hesitation in making a very overt display of the latter set of actions is not really surprising. In any case, nuclear strategy related activity in most nuclear weapon states occurs behind thick veils of secrecy. This, in fact, is believed to enhance deterrence, particularly when nations have small arsenals. In this respect, then, India is behaving no differently from others. The other reason that holds India back from revealing much on its procedural and organisational details is the long held belief that nuclear weapons are essentially tools of political deterrence and not military instruments of war-fighting. Hence, there is a tendency in the political establishment to play down the requirement of making plans for possible use of the nuclear weapon in military conflict. Moreover, the fact that the military was, in any case, not an active player in nuclear decision-making, especially prior to 1998, added to this propensity.

The focus of India, therefore, in planning its national security with nuclear weapons, has largely been on creating circumstances that do not allow deterrence to break down (which is a good endeavour), rather than on planning operations in case of the unthinkable eventuality of deterrence breakdown (which, however, is not exactly helpful in enhancing deterrence). In fact, there is an inherent paradox of nuclear deterrence — it is meant to be a situation that would avert the possibility of nuclear war; but to ensure this outcome, a certain display of preparedness to use the weapon in military operations is necessary. The more this preparedness is evident to the adversary, the more credible, stable and sustainable would deterrence be.

Therefore, national nuclear strategy must strike a balance between *maintaining secrecy for enhancing deterrence* while simultaneously *allowing transparency, also, for enhancing deterrence*. There is a distinct psychological aspect of deterrence, and a mastery of the sophistication necessary to play this mind game is as important as the tangible technological, organisational or operational capabilities that underpin it. In fact, in the case of nuclear strategy, some developments *must be transparent* because the greater purpose is to signal *capability and resolve* to the other side. Surprise in military wherewithal and tactics to catch the enemy unawares is a virtue of conventional warfare, but secrecy over capability and resolve in the nuclear realm would prove self-defeating. Transparency, in fact, provides a modicum of reassurance to the adversary, and in the nuclear game, it is far better to have the enemy at ease, rather than on the edge, hounded by the fear of losing his arsenal if he did not use it first. To that extent, India's publicly announced doctrine performs the critical task of communicating the precise purpose of India's nuclear weapons and indicating the circumstances and manner of their employment.

At the same time, the doctrine also provides direction to the nature of nuclear capability that India must build. For instance, given the unique nature and, hence, the derived purpose of the nuclear weapon, the doctrine concludes that an acquisition of *credible minimum deterrence* (CMD) would suffice. While there is no definition of this in terms of numbers of warheads, it may be taken to mean a stockpile that would be sufficient to cause unacceptable damage to the adversary. This determination then demands an understating of the "unacceptability threshold" of the enemy. Some of the other parameters that impinge upon this calculation are the yield of the weapon, its reliability quotient, the chances of the survival of the entire range of retaliatory capabilities after a first strike, as well as the enemy's capability to defend itself. As long as a stockpile can meet these requirements, there would be little sense in stockpiling numbers in excess.

For instance, the nature of India's likely targets (mega cities with high density of population) negates the need for very high yield or even very many low yield weapons. The tests conducted in May 1998 are deemed to have provided India the capability to build weapons between 20–200 kilotons (kt). The trend internationally is coalescing around 120 kt as optimum yield weapons, which certainly is within India's capability. Even so, it must be highlighted that even the low yields of 15-20 kt that were used in Hiroshima and Nagasaki were of such high destructive nature that they have scarred the human mind enough to restrain such inhuman action since. This, when the nuclear weapons were dropped from a height that was less than optimum (or the damage to Hiroshima would have been far greater) and on a target not well chosen (since Nagasaki's topography saved it from a lot more destruction). Therefore, if the weapon is employed with far greater intelligence and clever planning, even the low yield, especially in the case of India's adversaries, could well cause damage that no civilised state could deem acceptable.

Nevertheless, the determination of CMD would change over time since it is not constructed in a vacuum. It must be credible with respect to an entity, in this case, a threat from an adversary. Therefore, as situations within the country and around it change, so would the calculation of the credibility of deterrence. For instance, as nations increase their technological capabilities of offence and defence or surveillance and interception, it would impact on the credibility of deterrence and call for necessary adjustments. Even though India's 'minimum' is unlikely to swing beyond 200-250 nuclear warheads in the foreseeable future, it is nevertheless imperative that nuclear doctrine, strategy, policy and action must exist as dynamic entities responsive to the myriad changes in international and regional nuclear trends, bilateral and multilateral relations, and technological capabilities.

It is with this in mind that the book provides pointers for India's march towards credible deterrence. The first three chapters of the book explain the trends in nuclear weapons thinking and strategies amongst the other nuclear weapon states, including

India's adversaries, to suggest how the country's nuclear strategy must respond to them. These chapters also clearly bring out the fact that unlike most other nuclear weapon states whose deterrence is not meant to apply against any specific threat (as it did during the Cold War years of East-West confrontation), classical nuclear deterrence against nuclear armed nations, Pakistan and China, is a concept that is still valid in the case of India. In fact, for most others, nuclear deterrence is actually irrelevant in the face of the predominant threat posed by non-state actors. Yet, they hold on to their arsenal as a *hedge against future uncertainties.* For India, its nuclear deterrent exists to cater for very specific threats from states with nuclear weapons.

To add to the complexities, both of India's nuclear adversaries enjoy a strategic 'all weather friendship' and a robust mutual nuclear and missile proliferation relationship. In fact, in the history of development of nuclear weapons, China stands out as the only country to have provided its strategic ally, Pakistan, with weapons design, material, as well as having conducted a nuclear test on its own soil on its behalf. Therefore, India's nuclear deterrence must credibly apply to the two nations not only individually but also project the capability of holding up in the rare instance of simultaneous or coordinated wars on both frontiers.

The next five chapters identify some of the critical components of credible nuclear deterrence – nuclear doctrine (Chapter 4), command and control (Chapter 5), challenges pertaining to the survivability of the nuclear arsenal (Chapter 6), ballistic missile defence issues (Chapter 7), and fissile material concerns (Chapter 8). The ninth chapter brings out the historical linkages between India's nuclear weapons and energy programmes and examines the implications of the opening up of international civilian nuclear commerce for both these programmes. The penultimate chapter highlights the complexities before India's nuclear strategy as it must grapple with the complex interlinkages of nuclear deterrence with conventional and sub-conventional warfare. In the case of the superpowers, there was an element of stability at the nuclear level and conventional wars were fought between their proxies who were not only non-nuclear, but whose wars never really threatened their territorial security or survival. India does not enjoy this luxury since the conduct of proxy war in this case occurs on Indian soil and raises the spectre of conventional war between two nuclear neighbours.

Stability at the nuclear level is, therefore, severely strained by instability at lower levels, which is too close for comfort. But, if that risk is to be completely obviated, then, as the last chapter argues, India must continue its efforts for a nuclear weapon free world since that holds the best promise of nuclear security. However, until good sense prevails upon the international community to realise such a world, India will have to continue to maintain a credible nuclear deterrent. The book is written with the hope of helping India in this journey and thereby taking the country and the region far from the possibility of a nuclear holocaust.

Books are a labour of love. But the sources of that labour and love are many and it is practically impossible to faithfully list all individuals or influences that may have contributed in some way or the other. Thoughts consciously imbibed or inadvertently triggered by people met, read or heard guide the hand as it punches the keyboard. So it was with me. I have knowingly borrowed or unconsciously partaken from all that came my way. Of course, frailties of argument or inadequacies in expression are all mine. But for all that may be worthwhile in this book, I gratefully acknowledge the help of many.

The largest chunk of my gratitude goes to Air Cmde Jasjit Singh (Retd), Director, Centre for Air Power Studies (CAPS), who constantly encouraged me to look over the hill. In my association of over a decade with him now, he has been my mentor and true *director* –

providing me with a vision and helping me achieve it. I hope some of his ability to simultaneously balance the bird's eye view from afar and the worm's eye view from within has rubbed off on me. Shri K. Subrahmanyam, India's pioneer nuclear strategist gave me freely of his time (despite his other preoccupations) and perceptive understanding of strategic issues. Several students of nuclear strategy claim him their *guru*, and since I have gained a lot from many of his students, it makes him my *grandguru*! Brig Gurmeet Kanwal (Retd) was a constant source of inspiration and always an accessible sounding board for bouncing off my ideas. Rear Adm Raja Menon (Retd), Amb G. Parthasarthy, Prof Rajesh Rajagopalan and Brig Arun Sahgal (Retd) brought new consciousness of issues through their critiques as discussants of the initial drafts of some chapters. Air Marshal Ajit Bhavnani (Retd), former C-in-C, Strategic Forces Command, helped me understand the working of the Indian nuclear command and control structures. Dr Srikanth Kondapalli shared his perspectives on China and helped me locate the right sources of study. I am also grateful to Dr Dean Wilkening, Director, Science Programme, Centre for International Security and Arms Control, Stanford University, for his incisive inputs for the draft chapter on missile defence.

My colleagues at CAPS kept me in good cheer through the days of toil. I must particularly mention Air Vice Marshal Kapil Kak (Retd), Additional Director, CAPS, whose kind and encouraging words for all my work – papers at seminars or articles in the media — kept me going. Shalini Chawla provided a helpful ear to vent my occasional bouts of frustration and her practical advice and optimism egged me on. The staff in the library, Viji, Surbhi and Gautam, were ever ready to dig out or source books and papers that I requested. Also my gratefulness to the CAPS pantry staff for keeping me supplied with the necessary cups of tea so critical in an endeavour like this. My sincere appreciation for the painstaking and absolutely professional editorial effort put in by Ms Rehana Mishra. Jose, in the publishing house, finally turned the A-4 sheets of manuscript into a book and I thank him and Knowledge World for the thrill of seeing it in print.

Finally, this work would never have seen the light of the day if I did not have the loving family support that I am blessed with. My husband, Amulpreet, has been my biggest critic, and fan. Extremely proud of my work, he has always prodded me to seek excellence and been at my side to shoulder responsibilities of hearth and share matters of heart. My father never wavered in his implicit faith in my capabilities and helped me climb every step. My father-in-law shared his experience of thirty-five years at the Nuclear Power Corporation (NPCIL) which helped me understand the technical as well as management issues of the Indian nuclear power programme. My mother and my mother-in-law were ever ready to lend a helping hand at home when the call of duty beckoned. Their blessings, support and motivation have carried me through. Last, but not the least, a big thank you to my son, Keerat, for understanding that there were days when his mother just wanted to be left alone with the computer, and for his solicited and unsolicited youthful and uninhibited insights on profound issues.

It is to this easy exuberance of his and his generation, and to their faith in their own capabilities to tide over challenges, that I dedicate this book.

New Delhi **Manpreet Sethi**

Nuclear Defence Through Deterrence

Allow me to begin with an anecdotal experience. Soon after the nuclear tests in May 1998, the Japanese government set up a task force called the "Tokyo Forum" comprising 16 international experts on nuclear issues to come up with recommendations against clearly spelt out terms of reference which sought ways and means of getting India and Pakistan to give up their nuclear weapons and to work out future non-proliferations measures. The evening reception was hosted by Prime Minister Obuchi who addressed this high-powered group from across the world to exhort them to find a desirable outcome from the Japanese point of view. When I was introduced to him (then as Director of IDSA), he politely mentioned that the Japanese people were greatly upset and he hoped that we in India were conscious of their sentiments. My instinctive and spontaneous response was to assure him that indeed we were, and I personally had paid my respects at the

---

* Air Commodore **Jasjit Singh**, AVSM VrC VM, recipient of the *Padma Bhushan* for a life-time's contribution to national defence and security, is the former Director of India's premier think tank, the Institute for Defence Studies and Analyses (IDSA). He has authored many books, including *Air Power in Modern Warfare, India's Defence Spending, Defence from the Skies,* and *The ICON* (biography of Marshal of the IAF Arjan Singh), etc.

Hiroshima memorial twice in previous years. I added, "The Indian people, Mr. Prime Minister, who have had deep sensitivity and felt anguish about the catastrophic tragedy that the Japanese people had suffered, do not wish to see a Hiroshima in India." And, hence, the nuclearisation, since we were left with no other viable option to defend ourselves from such a threat. I am relating this incident to emphasise my own thinking on the purpose of our nuclear weapons and, hence, its doctrine and strategy.

Obviously India's core security interests required that it would – and must — do its utmost to ensure that a nuclear weapon is not dropped on its territory by a hostile power and nor should a threat of doing so lead to success in blackmail. In other words, nuclear weapons should not impinge on our foreign and security policies. There are three ways in which it can be ensured. First, abolish nuclear weapons from the inventories of nations. This, of course, is promised in Article VI of the Nuclear Non-Proliferation Treaty (NPT) – a treaty which has been praised by the P-5 and their allies for four decades. But even a cursory reading of facts indicates that those who happened to possess nuclear weapons before January 1967 have retained the right to not only continue with that possession, but actually increased the number of weapons to enormous levels (the US alone had 32,000 nuclear weapons by 1967). Worse, the contempt they have displayed over the past four decades in not even moving towards the implementation of Article VI of the NPT indicates their deep rooted belief in maintaining their hold over nuclear weapons while exploiting the rest of the NPT to ensure that others do not acquire similar capabilities.

While there has been recently an apparent move toward global nuclear disarmament (and even President Obama has lent public support to the idea), it is not clear whether this would ever reach the desirable end-state of a non-nuclear world. Under these circumstances, India cannot accept unilateral disarmament in the face of two of its not so friendly neighbours possessing nuclear weapons. In fact, Pakistan's declaratory policies relying on a "first use" doctrine and some of its actions (like holding out nuclear threats more than once since it acquired the capability by end 1986) hardly leave any scope for unilateral disarmament.

On the other hand, for India, *total abolition of nuclear weapons in reality is a core national security interest*; and so is not signing the NPT since that would force the country to become a non-nuclear weapon state facing two nuclear weapon states with whom we have significant territorial disputes and who have not hesitated to launch unprovoked military aggression against us in the past. India has been one of the leading members of the international community to support and advocate global nuclear disarmament. In 1998, however, it carried out five nuclear tests and also declared itself as a nuclear weapon state, in the process providing the requisite capability for credible deterrence. But it also committed itself to keeping it at the very minimum level so that precious resources required for the upliftment and well-being of hundreds of million of people should not be frittered away in search of nuclear ghosts or trying to be a Rambo.

Nuclear deterrence, to be effective, must be credible. This requires a strategy that is able to generate a great deal of synergy between at least five pillars of 'C's:

- **Clear aim** as to the purpose of nuclear weapons and what to expect from them. In our case, this can only be to deter the threat and use of nuclear weapons – and nuclear weapons only – against India and its forces. This requires to be stated and understood unambiguously right across the nation, especially by its civil and military elites; and any diplomatic advantage or "prestige", however small, that may accrue from being a nuclear power should be treated as a fringe benefit to be used to moderate the nuclear challenges till we reach the desirable end-state of a non-nuclear world.

- **Credible deterrent** which requires a rationally argued, objective doctrine and strategy to achieve the aim specified above. Unfortunately, this seems to have been diffused somewhat by the January 2003 official statement of the doctrine linking it up with use of large scale chemical and biological weapons. The probability of such large scale use of chemical and biological weapons is much higher by non-state actors (no doubt supported and supplied clandestinely by some state or the other) and there would be serious undermining of the nuclear deterrent when it is unable to respond to such attacks.
- **Capability** to get to the centre of gravity of the aggressor and deliver the counter-attack with nuclear weapons is critical for the credibility and effectiveness of the strategy. This calls for delivery systems and a sound targeting strategy which are both still works in progress.
- **Consensus or common understanding** (at least among a majority of the elites) of the nuclear strategy being adopted so that last minute hesitancy or debate/lack of understanding about what to do should not stymie the "assured retaliation" spelt out in the doctrine. But here we also need to rethink the official position articulated in January 2003 that altered the assured retaliation to "massive retaliation" undermining the credibility of the doctrine and, hence, the strategy of deterrence.
- **Coherence**. Above all, there is a vital need for (national) coherence in words and actions to achieve the ultimate effect. This is not easy in a democracy, especially where the understanding of the issue in individual minds may be limited or even absent when it comes to nuclear strategy. Unfortunately, even the Standing Committee on Defence of the Parliament

has shied away from discussing this subject; and many of our political leaders have a propensity to go in for rhetoric to impress their constituencies, and the media happily repeats it without questions. This has the potential of (mis)leading the adversary to take wrong decisions which could be disastrous for both sides.

When the concept of disarmament was included in the nuclear doctrine document of the National Security Advisory Board in 1999 and made public soon after, there was a lot of discomfort in many minds as to why this was being included in what some of them thought was already a weak and defensive strategy. The reason as stated above was that this was part of the principles and concepts (and, hence, doctrine) related to nuclear weapons. But the over-riding reason was to highlight the reality that global nuclear disarmament was, is, and will remain, a national security interest. Nuclear weapons, though critical to possess for nuclear defence when our security environment is nuclearised, actually complicate our security environment. Pakistan obviously would have remained less capable in conventional military capabilities which, however, did not stop it from launching wars against us; but it is also clear that it would lose those wars, as it did in the past. And that triggered its passion to acquire nuclear weapons to "neutralise" India's conventional military superiority which its foreign minister once described as the Sword of Damocles hanging over Pakistan's head.

Pakistan had started all its aggressions with irregular forces led by its military leaders to create internal turmoil in Jammu and Kashmir (J&K) and thus tie down large numbers of forces which would reduce the quantum that could be fielded against its own military. This covert war was followed up with overt armoured offensive in the hope that the Indian military may fall between the two stools of covert and overt warfare. But the point we often forget to note is that India did not hesitate each time in taking prompt and robust action even accross the Line of Control in J&K or even the international border in Punjab and Rajasthan to dispose of the covert irregular war and also defeat the regular offensive. But since 1987, by which year Pakistan had acquired nuclear weapons and concurrently started the covert war in J&K, there appears to have been a hiatus in our responses, limiting them to just killing terrorists within our own territory. The apparent reason, clearly interpreted as such by Pakistan, especially after the Operation Parakaram mobilisation in 2001-02, has been that Pakistan's nuclear weapons have deterred India from a punitive conventional response in the sanctuary from where *jehadi* terrorism has been prosecuted. And with an unlimited supply of poor unemployed *jehadi* youth motivated in many *madrassas*, mere killing of terrorists in India would not change policies in Islamabad (or, I should say more accurately, Rawalpindi which houses the GHQ of the army and also its "covert forces command" — the Inter-Services Intelligence).

One can take some solace in the fact that the potential nuclear equation with China is likely to be more stable, if for no other reason than that the doctrines of both countries are similar. And as long as they stick to that doctrine and strategy of no first use, the risks of nuclear weapons reduce enormously. On the other hand, having accepted the principle of no first use, both China and India are eminently placed to press for drastic reduction of nuclear weapons worldwide, shifting the doctrine to no first use and corresponding counter-strike strategies, to prepare the grounds for total global abolition of nuclear weapons. China may have legitimate concerns of American superiority in conventional military capability, especially when perceived in the context of past history and issues of Taiwan (or even Tibet). But it is already racing ahead of other Asian countries like India in a range of high-technology capabilities like space, information technology, greater reliance on conventionally armed ballistic missiles and cruise missiles for "beyond contact warfare," etc. which should provide it the confidence to defend itself, sooner rather than later.

From our own perspective, we need to remember that while China is occupying a large part of India in Ladakh region and claims the whole of Arunachal Pradesh, it would be almost impossible for it to take any territory by force across the rough line now being held by both. The solution of these territorial disputes is going to take a very long time, perhaps more than half a century from now; meanwhile, we just have to live with them like the Chinese would, since no serious stakes are involved which could provide the incentives for Beijing to reach out for either a solution or a conflict. We did lose the war in 1962, especially in one sector in the northeast and this weighs heavily on the Indian mind. But, the 1962 War was an aberration and cannot be replicated by China, especially as long as we take adequate and prudent precautions to ensure that our operationally deployable conventional capabilities are adequate for defence.

There is an obvious risk that given the strategic nexus between China and Pakistan, the two might decide to apply coordinated military pressure. But that does not necessarily imply a two-front war. Hence, what we need to watch closely is the two-front military modernisation that is currently underway to our north (helped by Russia) and west (assisted strongly by the United States, besides China). This requires giving greater priority to our own military modernisation which, if we see the unplanned depletion of the Indian Air Force's (IAF's) combat strength in the past six years because of decisions not taken in time (which has further tilted the air power balance against us), points to the problems that we might create for ourselves. This could potentially increase the risk of reliance on nuclear weapons in future, if the situation is not rectified. And while deterrence may work in extremis, the uncertainties of human behaviour under such circumstances cannot be relied upon for national defence of a country like India which itself is a rising power, and must have the ability to protect itself and its interests based on its own capabilities. And, hence, the need for greater attention to possession of conventional high-technology military capabilities.

Second, if global abolition of nuclear weapons is not feasible in the foreseeable future, then the option for Indian strategy is to ensure that military conflict, if it were to take place, is kept below the perceived nuclear threshold of the adversary. It is obvious that the aggressor's threshold is very difficult to assess accurately; and therefore, the need will arise to err on the side of caution to avoid any risk. But, this then severely constrains our ability to undertake punitive response to violent acts (especially increasingly virulent terrorist attacks over the past quarter century). This, in essence, has weakened, if not actually eroded our conventional deterrence to respond to sub-conventional war being waged by Pakistan under the rubric of "bleeding India through a thousand cuts." Hence, the need to understand the conflict environment facing us and to evolve a suitable strategy for punitive response to the war through terrorism under the nuclear umbrella that Pakistan has been waging for over two decades in the northwest, and China may be tempted to in the north-east.

The third, and most important choice is that till such time nuclear weapons are not surely and confidently abolished in the world, credible nuclear deterrence would have to remain the pillar of our nuclear defence, especially in the obviously uncertain and ambiguous world of nuclear weapons. This itself provides two options: to base the deterrence on a first use doctrine, or to rely on a counter-strike in retaliation to the adversary's first strike. India, in choosing not to be the first to use nuclear weapons, has opted for the counter-strike strategy with the categorical resolve and assurance that in case of an attack on India or its forces, there shall be assured retaliation "leading to unacceptable (for the adversary) damage." India, predictably and correctly, has opted for the latter doctrine for a variety of reasons. What is required is that a nuclear strategy should grow out of this doctrine to provide a credible deterrent against a nuclear threat or use.

Countries that possess nuclear weapons or aspire to do so have varied motivations for doing so and, hence, varying doctrines and strategies for their employment. Obviously, the superpowers were grossly in error in building stockpiles of over 60,000 nuclear warheads, which in the 1980s was claimed to be able to destroy each other under the infamous MAD (Mutual Assured Destruction) doctrine some 60-times over. But we die only once! As the Cold War intensified after 1978 in what was termed as the "Second Cold War" the build-up and deployments of nuclear arsenals of the superpowers became even more threatening and less controllable. In turn, the rising existential threat felt by people, especially in Europe and, Americas the also led to strong peace movements agitating against nuclear weapons. Even the UN felt the need to hold the Third Special Session on Disarmament in June 1988 where Prime Minister Rajiv Gandhi proposed a comprehensive plan to move toward a non-nuclear, non-violent world.

But soon after the Berlin Wall came down and the Cold War dissipated, it appeared that all the efforts to get rid of nuclear weapons would get a serious impetus. After all, nuclear weapons (with highly destabilising strategies of first use, etc.) were being justified as necessary because there was a Cold War. On the other hand, we found that the argument was turned on its head and the justification now put forward was that nuclear weapons were needed because there was no Cold War (!) and had been replaced with uncertainty; and nuclear weapons were required to take care of those uncertainties, virtually making out the case of a future self-fulfilling prophecy. It is in this background that Obama has come to power and promptly indicated a strategy for enhanced focus on disarmament and non-proliferation, which is believed to be on a tipping point.

Unfortunately, like the term deterrence, disarmament also means different things to different people. But for us, having adopted a minimum deterrence strategy for other good reasons, the real meaning is to go down to zero and institute sufficient measures to ensure that there is no future AQ Khan to upset that status. We are witnessing what may be termed as the third nuclear wave where a widespread set of former policy-makers under whom nuclear weapons acquired extraordinary numbers and legitimacy, today are arguing for abolition and/or disarmament. Obviously, the road to zero is going to be long and full of tangential centrifugal pressures. But it is a road worth taking, if for no other reason than that its time has come. The alternate may well be the collapse of the existing non-proliferation regime which reached its apogee a long time ago and can only absorb some tinkering like the Proliferation Security Initiative (PSI), etc. The grand bargains of nuclear arms control are coming to their end and either new ones would need to be negotiated, bringing the numbers substantively lower than generally assumed or nuclear anarchy may set in. It is worth recalling that based on empirical studies, Robert McNamara had concluded that 200 US weapons could destroy 25 per cent of the Soviet population and 50 per cent of its industrial base, making the state totally dysfunctional. And Russia is much smaller than the then Soviet Union, with its population expected to reduce by 17 per cent by 2050!

The major powers led by the United States had been working on inventing defence against nuclear attack that would reinforce the deterrent strategy; and, hence, the Anti-Ballistic Missile (ABM) Treaty of yore and the missile defence systems of current times. But what people forget is that the Ballistic Missile Defence (BMD) system, if it works for those who can afford the cost, would only provide defence against ballistic missiles as a delivery system. Other delivery systems already exist and for neighbours even a bullock cart may do! At the same time, the missile scene has been greatly complicated by China opting for very large numbers of ballistic missiles (over a 1,000 are reportedly deployed on the Taiwan front) presumably armed with conventional warheads as a force within the Second Artillery which controls and operates China's nuclear strategic capabilities. The bad news is that this would increase nuclear ambiguities and uncertainties where miscalculation risks would increase tremendously. On the other hand, the good news is that this would push countries to adopt the doctrine of no first use because of the heightened risks of miscalculation and errors of judgement, which could trigger an unwanted nuclear war.

At the same time, while the above complications are growing, we are witness to a tremendous revival of nuclear energy across the world which itself would have a profound impact on the ability of non-nuclear countries to acquire nuclear weapons. The rules of nuclear cooperation would have to be rewritten to ensure a safer world which is also sustainable by a non-conventional, renewable, environment friendly energy resource base that could last for the coming centuries. This, in a way, is the underpinning of the Indo-US nuclear deal; and we may see more innovative approaches in future, reconciling the challenges of nuclear proliferation with those of energy security for sustainable development across the world.

It is in this context that Dr. Manpreet Sethi has produced an outstanding treatise based on empirical facts and objective analyses covering nuclear strategy in a comprehensive manner not attempted by a scholar so far. I have felt privileged to have been asked to write my thoughts on some of the issues, though I am conscious that they are random, perhaps like nuclear strategy itself! This volume becomes all the more important because it comes at a time when a new interest in looking at nuclear weapons is becoming fashionable, even among the old Cold warriors; and it begins to shake the hiatus into which we seemed to have slipped after declaring ourselves as a nuclear weapon state more than a decade ago – the very period when we should have spent more effort and intellect to address the issues that the country will have to face in the coming years as a nuclear weapon state which would rely heavily on nuclear power and nuclear deterrence for defence while promoting abolition of nuclear weapons.

Jasjit Singh

List of Acronyms

ABM	Anti-Ballistic Missile
AEC	Atomic Energy Commission
ASAT	Anti-Satellite Weapon
ASW	Anti-Submarine Warfare
ATV	Advanced Technology Vehicle
BARC	Bhabha Atomic Research Centre
BAS	Bulletin of Atomic Scientists
BMD	Ballistic Missile Defence
BTWC	Biological and Toxic Weapons Convention
BWR	Boiling Water Reactor
CCS	Cabinet Committee on Security
CD	Conference on Disarmament
CDS	Chief of Defence Staff
CFE	Conventional Forces in Europe Treaty
CIRUS	Canada-Indian Reactor United States
CMD	Credible Minimum Deterrence
CNP	Comprehensive National Power
CONPLAN	Contingency Plan
CPGS	Conventional Prompt Global Strike
CSI	Container Security Initiative
CTBT	Comprehensive Test Ban Treaty
C4I2	Command, Control, Communications, Computing, Information and Intelligence
CWC	Chemical Weapons Convention
EMP	Electro-magnetic Pulse
DARPA	Defence Advanced Research Projects Agency
DRDO	Defence Research and Development Organisation
FAE	Fuel Air Explosives
FAS	Federation of American Scientists
FBR	Fast Breeder Reactor
FMCT	Fissile Material Cut-off Treaty

HEU	Highly Enriched Uranium
IAEA	International Atomic Energy Agency
ICBM	Intercontinental Ballistic Missile
INF	Intermediate Nuclear Forces Treaty
KANUPP	Karachi Nuclear Power Plant
KRL	Khan Research Laboratories
LOW	Launch-on-warning
LRTR	Long Range Tracking Radar
LUA	Launch-under-attack
MAD	Mutual Assured Destruction
MDA	Missile Defence Agency
MIRV	Multiple Independently Targeted Re-entry Vehicle
MRBM	Medium Range Ballistic Missile
NAPS	Narora Atomic Power Station
NATO	North Atlantic Treaty Organisation
NCA	National Command Authority
NCP	National Command Post
NFU	No First Use
NNSA	National Nuclear Security Administration
NPR	Nuclear Posture Review
NSAB	National Security Advisory Board
NSG	Nuclear Suppliers Group
NSSP	Next Steps in Strategic Partnership
NTM	National Technical Means
NWFW	Nuclear Weapon Free World
NWS	Nuclear Weapon State
NNWS	Non-Nuclear Weapon State
OSD	Office of the Secretary of Defence
PAEC	Pakistan Atomic Energy Commission
PAL	Permissive Action Links
PFBR	Prototype Fast Breeder Reactor
PHWR	Pressurised Heavy Water Reactor
PINSTECH	Pakistan Institute of Nuclear Science and Technology
PLA	People's Liberation Army
PLF	Plant Load Factor
PNRA	Pakistan Nuclear Regulatory Authority
PSI	Proliferation Security Initiative
QDR	Quadrennial Defence Review
RAPS	Rajasthan Atomic Power Station
RDD	Radiological Dispersal Device

RG Pu	Reactor Grade Plutonium
RMA	Revolution in Military Affairs
RNEP	Robust Nuclear Earth Penetrator
RRW	Reliable Replacement Warhead
SAC	Second Artillery Corps
SAM	Surface-to-Air Missile
SIOP	Single Integrated Operations Plan
SLBM	Submarine Launched Ballistic Missile
SORT	Strategic Offensive Reduction Treaty
SPD	Strategic Plans Division
SRBM	Short Range Ballistic Missile
SRF	Strategic Rocket Forces
SSMP	Stockpile Stewardship and Management Programme
SSBN	Nuclear-propelled Ballistic Missile Submarine
START	Strategic Arms Reduction Treaty
STRATCOM	Strategic Command
TAPS	Tarapur Atomic Power Station
TMD	Theatre Missile Defence
TNW	Tactical Nuclear Weapons
UNSCR	United Nations Security Council Resolution
WGPu	Weapon Grade Plutonium
WMD	Weapons of Mass Destruction
WPND	White Paper on National Defence

1

More of the Same
Nuclear Strategies of Major Powers and Implications for India

The year 2001 was significant for more reasons than just being the first year of a new millennium. It also marked ten years of the end of the Cold War and reflected a perceptible change of mood in international relations from the heady days of hope and optimism of the early 1990s.[1] Fast developing cooperative relations between the USA and Russia then, after nearly five decades of a bitter rivalry that was often perched on the edge of a cliff, had made officials and analysts euphoric. Some international strategists even dared to write obituaries of not just nuclear

1. Much of this optimism emanated from a reassessment of threats on the part of US and Russia. The predominant view among American security analysts of the time was that Russia lacked the capacity or even the intention to threaten or attack Western territories. Then CIA Director Robert Gates even told the US Congress in 1992 that "the threat to the US of deliberate attack from the [former Soviet Union] has all but disappeared for the foreseeable future." The then Russian Foreign Minister echoed the same sentiments when he too stated in the same year that his nation "no longer views the US as a foe." Tom A. Zamora, "Put A Safety Cap on Testing," *Bulletin of Atomic Scientists,* vol. 48, no.2, March 1992, p.25

weapons, but war in general.[2] One notable report published by the Stimson Centre recommended, "The ultimate objective of US national security policy should be the elimination of all weapons of mass destruction from all states – not the preservation of nuclear deterrence in perpetuity."[3] Taking a cue from the prevailing spirit of bonhomie, even the United National Security Council had concluded in a 1992 Summit Meeting, "The world now has the best chance of achieving international peace and security since the foundation of the United Nations."

By 2001, however, these expectations had come to naught as international relations developed along a completely different trajectory. It was on May 1, 2001, that President Bush made a landmark speech at the National Defence University, and given US leadership on most matters nuclear, this was significant. Rather dramatically, he declared, "Today, the sun comes up on a vastly different world. The wall is gone and so is the Soviet Union... Yet this is still a dangerous world, a less certain, a less predictable one...."[4] These words proved prophetic when only four months later, a singular event, the 9/11, transformed thinking on international security. Terrorism, possibly with weapons of mass destruction (WMD), facilitated

2. The spirit of the changing times was aptly caught by an international analyst in these words, "These are the days of hope , not despair... It is time to strip away the complex and arcane strategic theory of the Cold War and start from scratch." Lincoln Wolfenstein, "End Nuclear Addiction," *Bulletin of Atomic Scientists*, vol. 47, no. 4, May 1991. Also see William Arkin, Damian Durrant and Hans Kristensen, "Nuclear Weapons Headed for the Trash," *Bulletin of Atomic Scientists*, vol. 47, no. 10, December 1991, p. 16
3. Barry Blechman and Cathleen Fisher, "An American Legacy: Building a Nuclear-Weapon-Free-World," *The Final Report of the Steering Committee Project on Eliminating Weapons of Mass Destruction* (Washington D.C.: Henry L. Stimson Centre, March 1997)
4. President Bush's speech on missile defence delivered at the National Defence University (NDU), Washington, May 1, 2001. Available at http://www.acronym.org.uk

by proliferation of WMD material and delivery technologies to states of proliferation concern and/or non-state actors was identified as the predominant threat.

While a new cache of foreign policy tools and military strategies was crafted by the US to meet the emergent challenges, one concept of the old guard that was retained despite altered threat perceptions was nuclear deterrence. This continued to perform its traditional task of holding the peace between nuclear states, though the requirement appeared non-existent given the altered nature of their relationship. The US Nuclear Posture Review (NPR) of January 2001 maintained the need for nuclear weapons. This thread was picked up by the Russian military doctrine that followed in November 2003 and together they set the tone for others. And so, today, every nuclear weapon state (NWS) covets its nuclear arsenal and maintains it at the heart of national security.

This chapter is broadly divided into two sections. The first examines current trends in nuclear weapons thinking and strategies in four major NWS[5], namely the US, Russia, UK and France, on the basis of four parameters of analysis – nuclear deterrence; ballistic missile defence (BMD); nuclear arsenals and weapons infrastructure; and arms control and disarmament. The second section identifies the trends that can broadly be discerned from current developments and provides some indicative implications of these for India's nuclear strategy.

Examination of Current Developments

A country's thinking on nuclear weapons and the emphasis it places on them can be obtained from a number of indicators. For

5. Given the more direct implications of nuclear strategies and capabilities of China and Pakistan, they merit individual examination, which is undertaken in subsequent chapters.

the purpose of comparative analysis, this chapter has identified four parameters that in the assessment of the author best reveal the trends that the nuclear strategies of the four major powers appear to be following. The following sub-sections examine ongoing tangible initiatives (by way of deployments and/or research and development), as well as conceptual debates in the fields of nuclear deterrence, BMD, nuclear weapons infrastructure, and arms control and disarmament.

Nuclear Deterrence

Very early in the development of nuclear strategy, the primary purpose of nuclear weapons was identified as nuclear deterrence and so it has remained over the last six decades. From the US onwards, every state that has acquired nuclear weapons has justified them for their ability to deter another's use of nuclear weapons against itself. Nuclear deterrence has, therefore, remained at the heart of national security strategies of all nuclear weapon states. This trend has managed to outlive major upheavals in international relations. For instance, with the disintegration of the Soviet Union, many predicted the demise of nuclear deterrence too. In fact, with the significant transformation in US threat perceptions, even George Bush, Jr., during his election campaign in 2000, promised to leave the Cold War behind and "rethink the requirements of nuclear deterrence." While this was then interpreted as holding the promise of a gradual reduction in the role of nuclear weapons in US defence strategy, 9/11 and the emergence of two new threats—those arising from WMD proliferation to "rogue states" (in Bush parlance), and from the avowed interest of terrorist organisations in acquiring WMD—actually pushed the USA in the other direction.

US conceptualisation of a new strategic framework and the

role of nuclear weapons therein were articulated by the NPR[6] in 2001. It was premised on three assumptions:

1. The US must *retain a robust nuclear arsenal* because nuclear weapons provide a broad range of benefits to US defence and foreign policy. First, they provide deterrence against nuclear attack, and also against chemical and biological weapons. Second, they deter other equivalent powers from launching or winning a major conventional war. And, last, with some modifications, they could help in destroying targets hidden deep underground.
2. *Significant cuts in nuclear weapons are possible* because Russia is no longer the major threat confronting the US. The NPR, therefore, projected a scaling down of operationally deployed warheads to 3,800 by 2007, and their further reduction to 1,700-2,200 by 2012.
3. The size and composition of the US arsenal should be governed by the principle of "strategic adaptability"[7] to meet threats arising from multiple potential opponents and sources of conflict that may include rivals for global or regional leadership, 'rogue states', and terrorists. The United States should also prepare for the possibility of rapid escalation of existing threats or sudden emergence of new ones. This risk held back the NPR from recommending a complete destruction of dismantled weapons. Rather, it suggested that the downgraded weapons be moved as reserves in the nuclear force structure, where they could remain intact, and ready for recall if necessary. Termed

---

6. The NPR is a comprehensive review of US strategic nuclear policies, doctrine, force structure, command and control, operations, supporting infrastructure, safety, security, and arms control. The congressionally mandated report, prepared by the Secretary of Defence, also examines the selection of targets, stockpile levels, and new and potential threats to the United States and its allies and interests worldwide.
7. Term as used by Ivo H Daalder and James M Lindsay, "A New Agenda for Nuclear Weapons: On Nuclear Weapons, Destroy and Codify," *The Brookings Institution Policy Brief*, no. 94, 2002.

the Responsive Force, this would augment an already existing "hedge" of some 2,500 active and inactive nuclear weapons[8] and serve three purposes: act as replacement for warheads destroyed each year in routine reliability and safety tests; provide a safeguard against failure of operationally deployed weapons; and, enable a quick increase in size of the operational arsenal in the face of unexpected developments.

Based on the direction provided by the NPR, American nuclear deterrence rests on flexibility to quickly respond to the changing strategic, technical, operational and political environment. Five years after the NPR, the Quadrennial Defence Review (QDR) of 2006 too recommended a shift in emphasis from a "one size fits all" nuclear deterrence to a more "tailored deterrence" capable of handling threats from small powers, terrorist networks, as well as near peer contingencies. While classical nuclear deterrence was believed to be capable of managing the last, deterrence coupled with defence, counter-proliferation and preemption was considered necessary to provide security against small WMD capable nations, as well as against non-state actors. In fact, Ryan Henry, Deputy Undersecretary of Defence for policy had opined in 2005 that though deterrence of terrorists was most difficult, it was not impossible if the US could identify what they held dear and hold that at risk. Also, as these groups did not operate in a vacuum, it was possible to identify governments or territories that supported their operations and threaten them with preemption or retaliation.[9]

Therefore, in the face of new threat perceptions, nuclear deterrence is today considered one, but not the sole arrow in the

8. The present hedge comprises mainly retired warheads from the 1987 INF Treaty and the 1991 START I accord, which required destruction of delivery vehicles (bombers and missiles) but not the warheads.
9. As cited by Pamela Hess, "Bring Back the Nukes," http://www.spacewar.com, December 14, 2005.

US security strategy quiver. In a move towards operationalising the synergy of actions that would be required, the Department of Defence has been upgrading its nuclear strike plans to reflect a transition from the Single Integrated Operations Plan (SIOP) of the Cold War to smaller and more flexible strike plans for modern adversaries. For instance, one such plan, the CONPLAN 8022, developed by the STRATCOM (the Pentagon's strategic command) in coordination with the US Air Force and Navy provides "prompt global strike options to the President with nuclear, conventional, space and information warfare capabilities to destroy time urgent targets."[10]

An Interim Report made public on December 15, 2008, by the Congressional Commission on the Strategic Posture of the US provides the most recent endorsement for continuance of the centrality of nuclear deterrence for the US. Though the report recommends fewer nuclear weapons, it nonetheless maintains, "Given the uncertainties in the factors affecting global security today, *the need for deterrence (and extended deterrence) could extend for an indefinite future.*"[11]

If the US is in no hurry to reduce the role of nuclear weapons, neither is Russia. In fact, Russian nuclear muscle flexing in the post-Cold War era has largely coincided with the souring of the dream of a cooperative partnership with the West.[12] In December 1999,

---

10. Robert S. Norris & Hans M. Kristensen, "US Nuclear Forces, 2006," *Bulletin of Atomic Scientists*, vol. 62, no.1, January- February 2006.
11. Interim Report of the Congressional Commission on the Strategic Posture of the United States submitted pursuant to Section 1060 of the National Defence Authorisation Act for fiscal year 2009, December 15, 2008. p. 5. Emphasis added.
12. In the initial aftermath of the USSR break-up, President Yeltsin and Foreign Minister Andrei Kozyrev had sought a close relationship with the West. However, this tendency had dissipated by the end of the decade of the 1990s as a result of soured relations over the war in Bosnia, expansion of NATO, and growing Russian unhappiness over the American tendency to unilaterally use power.

during his visit to China, then President Yeltsin had underlined Russia's nuclear status when he said, "It seems Mr. Clinton has forgotten Russia is a great power *that possesses a nuclear arsenal*... I want to tell President Clinton that he alone cannot dictate how the world should live, work and play. It is us who will dictate."[13] Only four months later, in April 2000, his successor, Vladimir Putin again reminded the world that Russia is a "mighty nuclear power." Such statements reflected a need on the part of Russia to convince itself, and the world, that despite its economic disarray, political turmoil and reduced conventional might, it remained a power of international standing.

Russia's nuclear status is of crucial importance for the country at this present juncture for two main reasons. First and more obviously, its nuclear weapons fill an internal void by serving as an overt symbol of strength and power when other determinants of national status and international prestige have declined. They give Moscow the requisite strength to pursue domestic policies of the kind that it deems fit and to insulate it from external criticism on matters Russia considers its own, such as its policy on Chechnya.[14]

At the same time, the Russian nuclear arsenal also serves an external need. It fortifies an otherwise weak nation against the relative strength of some of the other international players

---

13. "Yeltsin Rattles Nuclear Sabre after Criticism over Chechnya," *Disarmament Diplomacy*, Issue no. 42. Emphasis added.
14. "Russia Calls NATO 'Cynical' Over Chechnya", *Reuters* (Moscow), December 3, 1999. It may be recalled that the first wave of "nuclearisation of Russian military thinking" surfaced after the end of the first war in Chechnya in 1997, when Russian Defence Minister Igor Sergeyev commissioned programmes for the development of the Topol M ICBM and the Iskander dual use tactical land based ballistic missiles. For more on this, see Alexander A. Pikayev, "Speculations on Russia's Deterrence Policy," in Ian Kenyon and John Simpson, eds., *Deterrence and the New Global Security Environment* (Oxon: Taylor and Francis, 2006), p. 102

that Moscow perceives as security concerns. For instance, the expansion of the North Atlantic Treaty Organisation (NATO) to areas traditionally considered a Russian preserve, as also the likely deployment of components of the US BMD architecture in these countries, or the increasing US unilateralist tendencies, are viewed with grave concern. Placed in such a security matrix, Russia perceives value in its nuclear weapons to underscore its status as a powerful nation.

Not surprisingly, therefore, over the last eight years, the Russian government has sought to reassert its nuclear strength through plans for new nuclear weapon systems, upgradation of existing ones, as well as the conduct of military exercises that contemplate the use of nuclear weapons. This process is premised on contemporary Russian security strategy that maintenance of a robust nuclear deterrent is critical for Russia. Fortunately for Moscow, the constraint of limited resources that until recently had prevented it from overhauling its old military equipment is now changing. Increased oil revenues allow greater defence spending, including on modernisation of the Strategic Rocket Forces, a trend that had stopped in 2000. This had been the result of the unhappiness of the conventional armed services over the 1997 policy that favoured nuclear forces over conventional modernisation, but which exposed the inadequacies in conventional capabilities in the Chechen War in 1999. In a reversal of policy of sorts, in the early years of the new millennium, Russian nuclear forces were largely ignored even though the military doctrine of April 21, 2000, had stated: "Under modern conditions, the Russian Federation relies on a need to possess nuclear capabilities enabling it a guaranteed possibility to inflict damage of predetermined scale to any aggressor under any circumstances."[15] It also mentioned Russia's right to use nuclear

15. Ibid., p. 107.

weapons in response to the use of not just nuclear weapons but of other WMD, or even in the case of conventional weapons. It may be recalled that Russia had abandoned its no first use (NFU) nuclear doctrine in 1992 in the wake of a dramatic decline in its conventional capabilities.

Nuclear deterrence remains central to Russian military and political strategy, and it is still largely conceived in the traditional sense of bipolar deterrence. Unlike the US, current Russian threat perceptions do not particularly revolve around non-state actors or WMD terrorism or even the need for deterrence of small nuclear nations. Russian fears still emanate from NATO expansionism and US unilateralism, both of which compel Moscow to sustain its nuclear deterrence capabilities.

Turning now to the United Kingdom, an overt debate on the need and role of atomic weapons that raged in British politics during 2006-07 was finally settled in favour of the country's retention of its nuclear arsenal. In October 2006, then Prime Minister Tony Blair had initiated a public debate on the future of the independent British nuclear deterrent. Two months later, in a White Paper entitled "The Future of UK's Strategic Deterrent," Blair underlined his government's belief that "an independent British nuclear deterrent is an essential part of our insurance against the uncertainties and risks of the future."[16]

After three months of deliberations, on March 14, 2007, the British Parliament approved the replacement of the country's submarine-based nuclear deterrent. This approval came in the face of significant opposition from within the ruling Labour Party, as also from non-governmental organisations (NGOs). The opponents argued that the UK did not need to replace the existing four Vanguard submarines and the Trident missile systems

16. Richard Cobbold, "Comment: A Debating Society," *RUSI Journal*, vol. 152, no.1, February 2007.

mounted on them and should instead live up to its commitments under the Nuclear Non-Proliferation Treaty (NPT) by scrapping nuclear weapons. But Blair favoured the move and reiterated his belief in nuclear deterrence as a key "insurance policy in an uncertain world,"[17] inhabited by states like North Korea and Iran, and by non-state terrorist organisations. He stated in the Parliament, "The government's judgment, on balance, is that though the Cold War is over, we cannot be certain in the decades ahead that a major nuclear threat to our strategic interest will not emerge."[18] Therefore, since the UK must maintain a credible nuclear deterrent, he argued for immediate action since the existing SSBNs (nuclear propelled ballistic missile submarines) would become obsolete by the mid-2020s and it would take until then to design, build and deploy new versions. However, he conceded the possibility of reducing the number of submarines from four to three, and the stockpiled warheads by 20 per cent from 200 or so to 160, in order to reduce potential replacement costs. With the decision now taken in favour of Trident replacement, the UK is set to retain its nuclear deterrent capabilities, even if in a scaled down version.

France appears to be in no different a mood. It was in 1960 that France acquired autonomous nuclear deterrence against Soviet nuclear and conventional superiority, and for the sake of its prestige and influence in Europe and beyond. Since then the *force de frappe* (nuclear strike force) has been central to French security and foreign policy. On January 19, 2006, then President Jacques Chirac had further reaffirmed this when in an address at the headquarters of the Strategic Air and Maritime Forces, he said nuclear weapons "continue, and will continue tomorrow,

17. "Blair Faces Backlash over Nuclear Deterrent Replacement," http://www.spacewar.com, December 3, 2006.
18. Lachlan Carmichael, "Blair Unveils Plans to Keep Nuclear Arsenal, Cut Warheads," http://www.spacewar.com, December 4, 2006.

to be the ultimate guarantor of our security." In fact, this speech was particularly remarkable for defining the French concept of nuclear deterrence in the post-Cold War world. Four points stood out[19]:

(a) It provided new missions of deterrence to nuclear weapons while retaining old ones. These included safeguarding broader French interests, ranging from territorial to strategic supplies (mainly energy); and deterring leaders of states that sponsor terrorism by hitting at their centres of power and capacity to act.

(b) It spoke of more flexible nuclear forces through enhancement of submarine and air launched nuclear weapons to address new situations and "cover threats wherever they arise and whatever their nature." Quite like the US, France expressed the need for quick and responsive nuclear forces that presented the option of decapitating a regime without causing civilian casualties. In fact, a report to the Senate Foreign Affairs Committee in 2003 sought an adaptation of nuclear deterrence from just "anti-city deterrence" to one that could provide a more variegated response through weapons with "selective yields and improved penetration of the nuclear warheads and a larger variety, enhanced range and higher precision of the carrier system."[20]

(c) It maintained that nuclear weapons give France "the ability to keep our freedom to act, to control our policies, to ensure the durability of our democratic values."

(d) It described nuclear weapons as a matter of national pride since it reflected the "marshalling of all energies, development of our research capabilities and finding innovative solutions...

19. "Chirac Reasserts French Nuclear Weapons Policy," *Disarmament Diplomacy*, no. 82, Spring 2006.

20. Dominique Lalanne, "New Nuclear Strategy, New Nuclear Weaponry," *INESAP Bulletin*, no. 23, 2003.

Nuclear deterrence, thus, became the very image of what our country is capable of producing."[21]

President Sarkozy too has maintained the centrality of nuclear weapons for his country's security. While inaugurating a new generation nuclear submarine of the *Triomphant* class in March 2008, he justified the need for his country's nuclear arsenal on the basis of the new range of threats, including from potential intercontinental ballistic missiles (ICBMs) from Iran. In fact, trying to provide the French nuclear weapons a wider role, he offered a dialogue with other members of the European Union to discuss the role of French nuclear weapons in Europe's collective defence. According to him, "Regarding Europe, it is a fact that France's nuclear forces, by their very existence, are a key element in its security. Let's together draw the logical conclusions: I propose to begin with those of our European partners who wish an open dialogue on the role of deterrence and its contribution to our common security."[22]

France, however, has declared that it would reduce the number of its warheads from 348 to 300. The reductions are envisaged in the air delivered weapons, even as the submarine-based weapons would stay in place, as these are described as "life insurance" against new threats. Overall then, the trend of retaining a nuclear deterrent capability that would be large enough to deter smaller states, as well as 'rogue actors' is obvious.

Ballistic Missile Defence

In his speech of May 2001, President Bush had stated, "Deterrence

21. Chirac's speech, n.19. The last two points were criticised for providing a rationale for nuclear weapons acquisition to other states, most notably Iran, for whom its nuclear programme has become more a symbol of status and pride.
22. Steven Erlanger, "Sarkozy Defends France's Nuclear Arsenal," *International Herald Tribune*, March 21, 2008.

can no longer be based solely on the threat of nuclear retaliation," and, hence, the need for "new concepts of deterrence that rely on both offensive and defensive forces."[23] Meanwhile, the NPR too had outlined defence (including missile defence, air defence, border security, etc.) as a crucial leg of the New Triad, the other two being the development of a combination of non-nuclear and nuclear strike capabilities to enhance deterrence; and, a revitalised defence and nuclear infrastructure. As explained by J.D. Crouch, US Assistant Secretary of Defence and official spokesperson for the NPR in 2001, the BMD aimed at creating a "broader array of capabilities... to deter and, if deterrence were to fail, to defend against potential adversaries."[24] It was described as an imperative in the new security environment in which risks emanated more from a few ballistic missiles of rogue states rather than the many in the Russian arsenal.[25]

Over the last at least a decade, the US has been engaged in aggressive research and development (R&D) as well as phased deployment of an active, integrated and multi-layered missile defence. Having abandoned the Anti-Ballistic Missile (ABM) Treaty in June 2002, the US BMD programme envisages interception of incoming enemy missiles at all stages of flight: boost phase interception to attack them soon after launch and while still over enemy territory; mid-course ground-based interception; as well as terminal stage interception. It is not within the scope of this chapter to recount the details of the three programmes. Suffice it to say that while these have seen their share of successes and failures over the last couple of years, the success rate has been

---

23. Bush's speech, n. 4.
24. As derived from the official briefing of the US Department of Defence, dated January 9, 2002, and made available by BASIC on http://www.basicint.org.
25. For more on rationale of BMD, see Manpreet Sethi, "Ballistic Missile Defence," in Jasjit Singh and Manpreet Sethi, *Nuclear Deterrence and Diplomacy* (New Delhi: Knowledge World, 2004) pp. 92-101.

comparatively higher in recent times. In fact, 2006-08 was an exceptionally good period for the BMD. Most tests were declared successful, engineering challenges were met and more nations signed on as a result of these successes, as also because of growing threat perceptions. The North Korean test in October 2006 pushed Japan and South Korea to reinforce their commitment to missile defence. The ongoing stalemate over Iranian nuclear intentions is pushing other countries in Europe to seriously consider the option. Meanwhile, in September 2006, NATO launched the Active Layered Theatre Ballistic Missile Defence Programme to protect troops in the field against short- and medium-range missiles by approving the construction of a $ 90 million command and control system. It is scheduled to have an initial operational capability by 2010 and become fully operational by 2016.

None of this, however, has been watched by any comfort by Russia. Moscow was always critical of US BMD plans.[26] But by the time the US left the ABM Treaty, it had become reconciled to the idea of an elementary missile defence, and crafted its own responses along two lines: a political response, such as rethinking the role of nuclear weapons and deterrence, scrapping the Strategic Arms Reduction Treaty (START) II, and more recently suspending the Conventional Forces in Europe (CFE) Treaty, as well as abandonment of the 1987 Intermediate Nuclear Forces (INF) Treaty which is widely considered the cornerstone of detente and peace and security in Europe. The Russian foreign minister indicated this in February 2007, when he said that Russia

26. A report entitled, "US-Russian Relations in Nuclear Arms Reductions: Current State and Prospects," prepared by the Russian Centre for Arms Control, Energy and Environmental Studies had stated in June 2001 that the US move was aimed at devaluing the Russian nuclear arsenal and warned that Russia would have to take adequate measures in response. As quoted in Eugene Miasnikov, "Russian Perceptions of US-Russian Security Relations and Responses to NMD," *INESAP Bulletin*, no. 18, 2001.

would stay within the INF only if, the US gave up its BMD plans in Eastern Europe; and indicated also a technological response, through modernisation of its nuclear forces and simultaneous development of BMD counter-measures. Perhaps it was a potent reflection of both responses that a Ministry of Defence Policy Paper of 2003 reemphasised the importance of the nuclear triad in Russian military strategy. Revising an earlier radical reduction of the land-based ICBMs in favour of naval submarine launched ballistic missiles (SLBMs), Russia indicated its decision to retain the former till the end of their service lives, including all extensions. It also hinted at maintaining the option of arming its Topol Ms (SS-27s) with multiple independently targeted reentry vehicles (MIRVs) as well as maintaining MIRVed ICBMs outlawed under START II.

By 2004, Russia had claimed to have developed three new long-range missile systems as part of its modernised strategic force. These included the mobile version of the Topol M ICBM, the sea-launched Bulava and a new, hypersonic manoeuvrable reentry vehicle to defeat BMD. In fact, the latter had been referred to several times by Russian presidents with immense pride from 2003 onwards. Significantly, in his state of the nation address in 2006, Putin indicated, "Work is already under way on creating… manoeuvrable combat units that'll have an unpredictable flight trajectory for the potential opponent."[27] While few details are available, these are believed to be hypervelocity scramjet systems capable of destroying targets at intercontinental range (10,000 km), with high accuracy and capability of a deep manoeuvre in altitude as well as direction. On counter-measures as well, Russia has worked to improve fast burn boosters, lofting or depressing ICBM trajectories, spoofing the interceptor missile's tracking

27. Sergei Kortunov, "Russia Must Stay Nuclear," http://www.spacewar.com, November 29, 2006.

sensors, and protecting the missile body with reflective or ablative coatings.

In keeping with the imperative of maintaining a robust nuclear deterrence, the budget for such programmes has also been increased. On February 7, 2007, Sergei Ivanov, Russia's Deputy Prime Minister and Defence Minister told the Duma that the Defence Ministry would receive $ 30.98 billion in 2007 as against $ 8.08 billion in 2001.[28] In September 2008, the Russian Prime Minister said that the country's defence spending would grow 27 per cent in 2009.[29] During 2007-15, the State Armament Programme estimates to spend $ 188.68 billion for a complete re-equipment of Russia's strategic forces. It plans to commission 34 silo-based missile launchers and command centres and 66 mobile Topol M ICBM systems, as also increase the number of strategic bombers, procurements that would largely go towards replenishing Russian nuclear forces that had degraded in their capability in the initial years of this decade. It is also planned that eight new strategic missile submarines would be built and commissioned, besides the development of space-based reconnaissance, telecommunications, data relay and early warning capabilities.

While Russia is working at maintaining its nuclear deterrent *vis-a-vis* a missile defended US, it has also vociferously expressed its disquiet over the US decision to deploy BMD elements – ten interceptors in Poland and a radar in Czechoslovakia – both former allies of the USSR. Russia perceives this move as having the potential to "raise tensions in Europe to a level they have not reached since the last great showdown in the Cold War…."[30]

28. As cited by Alexander Bogatyrov, "Russia Should Renew its Nuclear Arsenal," http://www.spacewar.com, February 13, 2007.
29. "Russia Test-Fires New Generation Strategic Missiles," http://www.spacewar.com, 18 September 2008.
30. Martin Sieff, "Rumblings From Russia," http://www.spacewar.com, October 19, 2006.

Russian analysts have argued that the real motivation behind this is "the desire to expand US military and strategic capacities and constrict those of other states that have nuclear missiles, Russia and China most of all," because even a limited BMD "injects a high degree of indeterminacy into the strategic plans of other countries and undermines the principle of mutual nuclear deterrence."[31]

The head of the Russian Defence Ministry's International Military Cooperation Department, Yevgeny Byzhinsky, described the American move as "unfriendly" and warned that Russia would be compelled to respond with "adequate retaliatory measures of military and political character."[32] Col Gen Boris Cheltsov, Chief of the Air Force has opined that deployment of ABM in Europe "will give the Pentagon and its NATO allies the potential to destroy Russian strategic nuclear forces at the most vulnerable stage: the initial ascending leg of the trajectory."[33] Though most Russian pronouncements dismiss the possibility of these deployments eroding its nuclear deterrence with the US, they do express two fears: one, of a possibility of the present rudimentary infrastructure leading to a build-up of military presence in Europe; and two, of the system acting as a lodestone for other countries such as Ukraine, Georgia or Moldova to invite the US to their territories. Not only would this bring US forces much closer to the Russian borders but also into countries that Moscow still considers fall in its area of influence. Therefore, Moscow is unconvinced by the US description of these moves as "limited preventive measures" and instead looks at them in the general context of the strategic and military-political environment and its relationship with the West, which is not going through the best of times. Moscow has

31. Mikhail Barbanov, "Why Russia Fears BMD," http://www.spacewar.com, February 15, 2007.
32. As cited by Sieff, n. 30.
33. Dmitry Danilov, "Europe's BMD Turmoil," *The Post Chronicle* (New York) April 12, 2007.

also expressed scepticism over American overtures for a joint missile defence system and has suggested instead a joint analysis of threats first.

While it is true that Russia is disturbed by US moves on BMD, it is equally true that several Russian developments of its strategic arsenal would have occurred even without the American provocation. US BMD overtures offer a ready justification to Russian and Chinese modernisation plans. Nevertheless, all of this does not bode well for international security. US-Russian relations are at low ebb at the start of 2009 and just two decades after the end of the Cold War. And, BMD is one, but not the only, reason for this.

Meanwhile, the UK and France are, for the moment, bit players in the BMD game. Both were initially sceptical of US moves in this direction. Their doubts arose from their uncertainty about the seriousness of the threat that the BMD sought to address, its technological feasibility, the financial cost of the exercise, the future of the ABM Treaty and international arms control, and the impact on relations with Russia and China. But once Bush's resolve to push ahead with the technology became clear, as also his desire to hold complete and close consultations with the allies on the issue, the two countries considerably toned down their rhetoric.

The subject has many facets for the two countries, given their membership of the European Union (EU) and NATO, both of which have different perspectives on missile defence. While the European Security and Foreign Policy of the EU has little focus on BMD or even a unified approach, NATO is engaged in developing a theatre missile defence (TMD) system. But, the US sees BMD as a national endeavour and seeks bases in Eastern Europe on the basis of bilateral agreements. In fact, the head of the US Missile Defence Agency (MDA) has expressed that US missile defence components could become a national contribution toward an alliance-wide defence system if and when NATO decides to establish such a system.

At the root of the differences between the US, EU and NATO positions has been a lack of unity on how much to rate the threat posed by Iran and how to deal with it. At one end was the position voiced by the former French President who had warned that "we must be very careful, as regards this project, not to encourage the creation of new dividing lines in Europe or the return to an obsolete order."[34] France is largely neutral on NATO deployments but is a stakeholder in European TMD efforts and is itself investing in early warning satellites. In fact, a news report of June 2008 indicated that France under President Sarkozy was likely to double its current budget for space-based BMD and other intelligence gathering systems in order to have an operational ground-based BMD system by 2015 and space-based surveillance and warning systems in orbit by 2020.[35]

On the other hand, the British government is involved in talks with Washington, already hosts a radar which feeds information to the US MD system, and is willing to discuss the potential deployment of interceptors in the UK. But the US itself is hesitant to take up this offer since missile defence bases on British territory would improve American ability to intercept Russian ICBMs, thereby raising more problems.

Nuclear Arsenals and Weapons Infrastructure

The health of nuclear weaponry (operational forces) and the infrastructure (development and production capabilities) provide a good indication of the nuclear thinking and strategies of a nation.

Contemporary US nuclear strategy has considered an upgrade of strategic offensive arms and the development of smaller nuclear

34. Chirac's remark during a press conference on March 9, 2007, and as cited by Oliver Meier, "Europeans Split Over US Missile Defence Plans," *Arms Control Today*, April 2007.
35. "Sarkozy Plans BMD Shield for France," *United Press International*, June 24, 2008.

munitions for possible use along with other smart weapons against hardened, underground facilities housing command centres or nuclear, biological, chemical (NBC) weapons related infrastructure. The NPR, in fact, had identified 1,400 underground sites worldwide that required targeting by the nuclear force because conventional weapons were inadequate to destroy them.[36] In order to perform this function, it recommended the development of a Robust Nuclear Earth Penetrator or an earth penetrating bunker buster mini nuclear weapon. The director of Sandia Laboratories too argued in 2001 that, "...the yields of weapons left over from the Cold War are too high for addressing the deterrence requirements of a *multi-polar, widely proliferated world. Without rectifying that situation, we would end up being self-deterred.*"[37] US military experience with Iraq, and the Taliban in Afghanistan who have effectively used caves as their hideouts, only strengthened the case for mini nukes.

For several months after this recommendation, debate raged over the feasibility and desirability of such a weapon. Supporters of these small, nuclear bunker busters argued that "precise" nuclear strikes with low yield warheads would minimise, if not obviate radioactive fallout and, hence, restrict collateral damage. This, however, is not true given that even a precise, low yield nuclear weapon cannot be expected to carry out as surgical a strike as is claimed. There are several imponderables that must be taken into account. Errors could arise as a result of slippages on pinpoint accuracy because of faulty target intelligence, problems with navigation or guidance systems, as well as reentry challenges. Dangers inherent in such an operation are, therefore, no less likely to lead to a catastrophe of unimaginable proportions. In fact, a report prepared for the

36. James J. Wirtz and Jeffrey A. Larsen, *Nuclear Transformation: The New US Nuclear Doctrine* (Hampshire: Palgrave Macmillan, 2005).
37. Robert W. Nelson, "Low-Yield Earth-Penetrating Nuclear Weapons," *FAS Public Interest Report,* January- February 2001. http://www.fas.org.

Federation of American Scientists (FAS) concludes,

> No earth-burrowing missile can penetrate deep enough into the earth to contain an explosion with a nuclear yield even as small as one percent of the 15 kiloton (kt) Hiroshima weapon. The explosion simply blows out a massive crater of radioactive dirt, which rains down on the local region with an especially intense and deadly fallout.[38]

In view of the deep divisions on the subject, specific approval for R&D of such a weapon was never forthcoming in the form of congressional funding.[39] However, the attraction of newer warheads never waned in the US. In 2005, the Reliable Replacement Warhead (RRW) programme was initiated when Congress allocated $9 million to explore making existing nuclear warheads last longer without diminishing their explosive power. This programme is euphemistically described as aiming "to improve the reliability, longevity and certifiability of existing weapons and their components,"[40] besides also enhancing their safety and security. It is also endowed with a more 'benign' purpose of providing research and engineering problems to the "unchallenged" nuclear workforce in laboratories that were also to be refurbished as part of the RRW programme. In fact, the National Nuclear Security Administration (NNSA) has argued that the programme would be useful for revitalising the existing "decrepit" nuclear infrastructure, which it claims had suffered under "decades of neglect, intense congressional scrutiny and legal enforcement actions in the late 1980s and 1990s." In an October 2006 report, the NNSA stated that a revamped

38. Ibid.
39. It may be mentioned that the Spratt-Furse Amendment of 1994 prohibited the US Administration to undertake R&D on low yield nuclear weapons, which were technically defined as those with a yield of less than 5 kilotons.
40. Jonathan Medalia, "Nuclear Weapons: The Reliable Replacement Warhead Program," *Congressional Research Service*, May 24, 2005, p.6.

complex would "improve the capability to design, develop, certify, and complete production of new or adapted warheads in the event of new military requirements."[41]

This refurbishment is the third leg of the new triad, and is expected to repair and replace nuclear weapons production and development facilities. Known as the Complex 2030 programme, for the year it is to be realised, it would enable the US to increase reliable warhead production capabilities while reducing actual numbers. In the process, it would achieve what Linton Brooks, Administrator of the NNSA, desired: not "a smaller Cold War era nuclear stockpile", but "capabilities appropriate for 21st century threats."[42] In his testimony before the Senate Armed Services Sub-Committee, Brooks said, "Today's Cold War legacy stockpile is the wrong stockpile from a number of perspectives"[43] because it contained too many high yield weapons that offered little or no capability against hard and deeply buried targets, provided no way to limit collateral damage, and could not destroy chemical or biological munitions. The RRW sought to rectify these limitations by matching nuclear weapons to the new threats. Moreover, in the biggest coup of arguments, it was claimed that the RRW would ensure security of warheads against physical attack by terrorists, thus, reducing security costs of manpower and equipment at nuclear facilities.

In order to make the RRW possible, the Department of Defence worked with the Department of Energy to finalise a new warhead design. Nuclear weapon laboratories were tasked to this effect and

---

41. Wade Boese, "New US Warhead Design Selected," *Arms Control Today*, April 2007.
42. As cited in Walter Pincus, "Plan to Study Nuclear Warheads Stirs Concern," *Washington Post*, April 6, 2006.
43. Statement as cited in Stephen I. Schwartz, "Warheads Aren't Forever," *Bulletin of Atomic Scientists*, vol 61, no. 5, September/October 2005. pp. 58-64.

in early 2006, Lawrence Livermore National Laboratory in California and the Los Alamos National Laboratory in New Mexico submitted competing designs for a new warhead. A period of review followed during which in January 2007, Bush even hinted at the possibility of a hybrid of the two designs to arrive at one that would be more robust and resistant to accidental or unauthorised use. But, in March 2007, the NNSA chose the Livermore laboratory design since it invoked greater confidence of being certified without nuclear testing. While announcing the decision on the new warhead design which will initially replace the W76 warhead for submarine-launched ballistic missiles, NNSA's Acting Administrator, Thomas D'Agostino said that the RRW programme would use a warhead design that was tested in the 1980s but package it with new features such as insensitive high explosives less liable to explode by accident, as well as locking devices that would prevent the warhead from being used if it fell into the hands of terrorists.[44]

While the selection of RRW seems to maintain the US moratorium on testing, it may not be forgotten that the NPR had reiterated opposition to American ratification of the Comprehensive Test Ban Treaty (CTBT), while recommending that the country should reduce the time necessary to prepare its nuclear test sites from the current period of two years to one year. This was despite the fact that the US has a Stockpile Stewardship and Management Programme (SSMP) that enables virtual nuclear testing to ensure continued viability of US nuclear weapons, train a new generation of weapon scientists, and take up refinement of weapon designs and production processes. Yet, the NPR had proposed accelerating test site readiness.

For the moment, it appears that international criticism may have been one of the reasons for the US selecting a new warhead

44. Walter Pincus, "Nuclear Warhead Plan Draws Opposition," *Washington Post*, March 4, 2007.

design that would need no new tests. Yet, it must be pointed out that the US Deputy Secretary of Energy, on a visit to Russia in early March 2007, soon after the selection of the new design, said that, the US reserved the right to conduct nuclear tests in the future, though he hoped those would not be necessary.[45] The head of the US Strategic Command, Gen James Cartwright also told the Senate Armed Services Strategic Forces Sub-Committee on March 28, 2007, that the RRW programme would not be terminated even if a need for testing was felt.[46] In fact, even though it is widely presumed that the new Democrat Administration of Barack Obama might consider resubmitting the CTBT to the Senate for ratification, the December 2008 Interim Report on the Strategic Posture of the US recommends that "before submission, the Department of Energy and Department of Defense should receive from the laboratories and STRATCOM clear statements describing the future capability and flexibility required to minimize the risks of maintaining a credible, safe and reliable nuclear deterrent without nuclear explosive testing."[47]

These developments are obviously not lost on Russia. Though there is not much information available on the exact work underway on nuclear weapons there, statements such as "Russia's strategic forces are the main priority on the national defence agenda," by none less than President Putin throughout his presidency convey the general trend. In mid-December 2006, Putin had called for the "development of cutting edge strategic weapons" and emphasised "quality over quantity."[48] Among some

---

45. "US Retains Right for Nuclear Tests if Necessary," http://www.spacewar.com, March 14, 2007.
46. John Fox, "Nuclear Testing Would be Debated, US General Says", *Global Security Newswire,* March 30, 2007.
47. Interim Report, n. 11, p. 11.
48. "Russia Against US MD Plans for Europe," http://www.spacewar.com, December 27, 2006

of the other major modernisation efforts underway in Russia, the launch of a new generation nuclear submarine in April 2007 deserves special mention not only because it seeks to upgrade Russia's undersea nuclear strike force, but because it is for the first time in 17 years that the country has launched such a vessel.[49]

With the decision on replacement of its SSBNs, the UK nuclear weapons infrastructure is naturally expected to be in a state of animation. Meanwhile, France too has been shoring up its nuclear deterrence capabilities. Having dismantled its land-based nuclear missile silos in 1996, the French nuclear deterrent is now based only on a dyad of submarines and aircraft. Hence, the French government has engaged in a more focussed approach towards modernisation of its nuclear forces. In November 2006, it conducted the test firing of its new 8,000 km range M-51 ballistic missile which is to carry submarine-based nuclear weapons. Expected to replace France's existing SLBMs by 2010, the new missile would be capable of carrying six nuclear warheads.

Arms Control and Disarmament

Given the current mood, as discussed in the above paragraphs, it should come as no surprise that arms control whether bilateral, (between the US and Russia) or multilateral [in the Conference on Disarmament (CD)] is a low priority. The CTBT has been in a state of limbo from 1999 onwards. In 2002, once the US walked out of the ABM Treaty, Russia announced its abandonment of START II. Recent Russian pronouncements on the INF Treaty have put it under a cloud and the Fissile Material Cut-off Treaty (FMCT) is still awaiting direction in the CD. In May 2006, there was a sudden spurt of activity when the US put forth a draft FMCT that other countries promised to study. But nothing substantive has occurred

49. "Hi-Tech N-Sub Launched by Russia After 17 years," *The Times of India*, April 16, 2007.

since then. All of these indicate a general trend away from arms control, particularly of the multilateral variety.

Instead, the US, normally the leader in these matters, has displayed greater inclination for unilateral initiatives to reduce redundant weapons while keeping the option of reversibility. This was evident in the negotiations and conclusion of the Strategic Offensive Reductions Treaty (SORT) in 2002. The brief (only 500 words long) US-Russia bilateral treaty codified cutbacks in a very loose and flexible fashion. It legally bound the countries to a reduction of their nuclear arsenals to a range between 1,700-2,200 by 2012 but allowed each side to determine the manner, sequence and timetable that it would follow and also how the dismantled warheads would be disposed off. The flexibility in this regard has obviously been maintained to uphold the US desire to reduce deployed weapons, but stockpile them in a reserve force. The treaty contains no provisions for limiting delivery vehicles, and in fact, allows the two countries to retain as many delivery vehicles as they like—without exceeding START I limits, which itself will remain in force until 2009 only, and expire before SORT reaches its targets.

SORT also does not include the elimination of MIRVed missiles, a step that START II had stipulated. This was obviously a concession to Russia since it deploys a greater proportion of its strategic warheads on MIRVed ICBMs and is likely to MIRV the new Topol M missiles too in response to the BMD. Neither have the USA and Russia laid down any verification procedures to establish the veracity of the cutbacks.[50] In fact, a move away from verification appears to be the current American tendency. The draft on the FMCT proposed by the USA in May 2006 at the CD also steers

---

50. The onerous and complicated task instead has been left to a Bilateral Implementation Commission that is expected to meet at least twice a year to deal with issues of compliance.

clear of enforcing any verification mechanism, leaving it instead to national technical means (NTM) to detect non-compliance.

Meanwhile, over 2007-08, there did emerge some strains of activity and interest in the universal elimination of nuclear weapons. Cold War practitioners George Shultz, Henry Kissinger, William Perry and Sam Nunn expressed themselves in favour of nuclear disarmament in two articles published in the *Wall Street Journal* in January 2007 and January 2008. Once responsible for crafting airtight nuclear deterrence doctrines for the US, the four stalwarts admitted the fallacy of the logic of nuclear deterrence and argued for reversing the direction of movement. Several other attempts are also evident elsewhere; for instance, a study to examine verification challenges of disarmament is being conducted in the UK, a Commission has been set up by Australia and Japan to examine the prospects of disarmament, and the Global Zero initiative is being led by Bruce Blair, a command and control expert for American weapons only some years ago.

The moot point, however, is the final objective, sincerity, and longevity of these efforts. For the time being, it does not appear as if the requisite critical mass that could propel the world towards elimination of nuclear weapons is anywhere close to being formulated. While political leaders everywhere make the right noises on the subject, everyone tends to place it as a long-term goal to be achieved some time, but not immediately. For the moment, therefore, nuclear weapons continue to be considered necessary to safeguard national security.

Manifest Trends and their Implication for India's Nuclear Strategy

Continued Salience of Nuclear Deterrence

While the world may have moved on from the centrality of the Cold War and its bilateral conception of nuclear deterrence, the underlying rationale of deterrence nevertheless remains valid and relevant. It is, of course, viewed differently and no longer straitjacketed only in mutual assured destruction (MAD). Rather, to cater for the new range of threats and uncertainties, it has been endowed with greater flexibility that allows a tailored response instead of a one size fits all approach. While classical deterrence prevails against major nuclear powers, it is enforced through additional instruments such as active and passive defence, counter-proliferation, and prevention of WMD proliferation against smaller arsenals of other states and against non-state actors. Deterrence, therefore, is no longer the sole keeper of security, but is one among an array of potential approaches.[51] There are several trends here that need to be considered in the Indian context.

Given India's threat perceptions, primarily from nuclear-armed neighbours with comparable or better nuclear forces, classical nuclear deterrence will have to remain an important pillar of national security for the foreseeable future. Therefore, India cannot afford complacency in the development of its credible nuclear deterrent capabilities. Meanwhile, though terrorism, including of the WMD kind, is perceived as a major threat, India, unlike the US does not seek to address it through counter-proliferation or preemption but with improved intelligence, export controls and better preparedness.

Secondly, given the current emphasis on flexibility in deterrence enforcement, it needs to be debated whether India too would need some changes in its nuclear doctrine to reflect this better. Right now, the doctrine is based on "massive retaliation". Does this restrict the choice of response and, hence, detract from credibility of deterrence? Would it be better to define the nature of retaliation more generically as "punitive", thus, allowing for a

51. A similar viewpoint has been expressed in Kenyon and Simpson, eds., n. 14, p. 9 ff.

more graded response? While it cannot be denied that the Indian nuclear doctrine is premised on deterrence by punishment and, hence, on counter-value targeting to hold cities at risk, it is nevertheless debatable whether massive retaliation would qualify for every contingency.

Thirdly, the NWS are seeking flexibility also in terms of the yield of nuclear warheads, mostly making them smaller, to meet different contingencies. For India, given its current capabilities, this is a non-issue since in any case it does not have the large megaton monsters of the Cold War period. It would be of far greater relevance for India to seek a range of delivery options through the triad – or the dispersion of nuclear forces on land-based missiles, sea-based missiles and air-delivered warheads. Diverse basing would provide greater freedom of deployment as well as enhance survivability, which is so critical with India's no first use strategy.

Deterrence with Defence

Unlike the situation in the Cold War when deterrence and defence were seen as mutually exclusive, the trend today indicates their cohabitation. This, however, means a change in the operation of deterrence from deterrence primarily by punishment, which was based on MAD, to the same also by denial.[52] This is expected to shore up deterrence by signalling to the enemy that not only would he suffer the consequences of initiating a nuclear war, but also face the prospect of losing the war while the other side is able to win it. Counter-proliferation, preemption and missile defence are the components of denial strategies and emphasise the war-fighting role of the nuclear weapon. However, this applies only in

52. For an informative analysis of this shift, see Rajesh Rajagopalan, "Nuclear Deterrence: The Return of the Denial Strategy," *Asian Strategic Review 1998-99* (New Delhi: IDSA, 1999), pp. 151-167.

the context of asymmetric threats from small WMD capable states or non-state actors. Therefore, the US can find it workable against small arsenals of the states of proliferation concern.

In the case of India, given the nature of nuclear and missile threats from Pakistan and China, deterrence by denial is largely irrelevant. This will be true even after the country has acquired a rudimentary BMD capability because it would at best only provide point or area defence.[53] Therefore, deterrence by punishment would have to be the basic premise for Indian nuclear strategy.

Reduction, but Refinement of Nuclear Weapons

All NWS have accepted reductions in their arsenals, whether in the number of warheads or delivery platforms, or both. But also, in all cases, these cutbacks are accompanied by moves towards refinement and modernisation of nuclear forces. The overall trend is generally towards a leaner and meaner nuclear force, including the development of small and more usable nuclear weapons.

As a country with a minimum nuclear deterrence strategy that is still engaged in building and operationalising a credible deterrent, the trend of reductions in stockpiles does not apply to India. But New Delhi must be attuned to the latest developments in the field of new nuclear warheads and delivery systems. It is also important that the country develops an adequate infrastructure that is capable of building and maintaining a credible force posture, and also that is sufficiently forward looking to provide new capabilities as required. While India obviously need not match the scale of nuclear weapons infrastructure of the larger powers, it must build it as a flexible and responsive one that can address uncertainties of the future. In fact, until now, India's nuclear programme has existed as a single programme for power production and weapons

---

53. For a detailed examination of the possible role and nature of BMD in India's nuclear strategy, see Chapter 7 of the book.

development. As a commitment under the Indo-US civilian nuclear cooperation, a Separation Plan was drawn up in March 2006 (See Appendix 4) to be executed once the entire process of the cooperation agreement had been completed. This will be undertaken in the coming years. Therefore, the formulation of an overall roadmap that lays down priorities and guides actions would go a long way in building necessary capabilities.

Moreover, India will have to take into account the tendency to project small nuclear weapons as more usable in some scenarios. India has always maintained that any use of the nuclear weapon would not only be a disaster, but more importantly also irreparably damage the moral taboo against nuclear use. This, then, could have serious repercussions for Indian security. For instance, Pakistan may interpret this development as it being acceptable to undertake a demonstration/small nuclear strike against India to check the advancement of conventional forces in case of a crisis[54]. In such case, if India responded with a nuclear strike, it would inevitably lead to a larger nuclear exchange that would be disastrous for both. But, if India did not respond, it would seriously undermine the credibility of its deterrence for the future. Therefore, it would be best if the nuclear weapon was not seen as a slightly more dangerous conventional weapon and never brought into use. It is in India's interest to urge a globally high nuclear threshold for all.

Possibility of Nuclear Testing

Though for the moment, the US has displayed a reluctance to conduct nuclear tests, it has not categorically ruled out such a move. Not only has the US not ratified the CTBT, American officials have hinted at the conduct of nuclear tests if required, as mentioned earlier in the chapter. Meanwhile, President Putin too

54. Interestingly, some Pakistani literature points to the same logic being used by India to mount a small nuclear attack against them.

had stated while visiting one of the Russian nuclear laboratories in 2003 that his country would refrain from nuclear testing "under certain obligatory conditions, one of the most important of which is a similar attitude on the part of other nuclear states..." Engaged as both are in a process of nuclear modernisation, and having zealously retained the option of underground nuclear testing, there is no guarantee that one or the other may not do so in response to a perceived need to restore strategic stability in an insecure environment. China's response in such a situation would have to be factored into India's nuclear deterrence.

India has declared a unilateral moratorium on testing, but it has refused to go beyond this commitment. While there is nothing to indicate that New Delhi would undertake a nuclear test without any provocation, it is loath to surrender the option not only because it provides insurance against future uncertainties, but also because an open option has symbolic value.

Unilateral Nuclear Cutbacks

On nuclear reductions, the general tendency seems to be in favour of parallel but unilateral reductions in the US and Russian nuclear arsenals without any mutual agreements or prior consultations. These cuts arise mostly from technical factors or economic expediency and are more by way of rationalisation of arsenals in the face of changed threat perceptions. Unilateral decisions on the composition and size of the national nuclear arsenal also help retain the option of reversibility since the reductions do not enforce any legal commitments.

Given this trend, India cannot count on China being involved in any multilateral nuclear reduction exercises led by other NWS, a possibility that has been referred to in the past. It was expected that bilateral nuclear arms control between the USA and Russia would eventually bring in other countries as arsenals dropped to their levels. But the present move is from bilateral to unilateral instead of towards multilateral controls. New Delhi shall, therefore, have to base the calculation of its own nuclear deterrent on current estimation and future projections of the Chinese arsenal, discounting the possibility of their coming down from present numbers, and not being surprised if they actually went up in response to the US BMD. China has expressed that it would maintain "pace not race" with the US, but it is determined to hold Washington to the deterrence equation. India too will have to maintain the same with China.

Dwindling Faith in Multilateral Treaties

A corollary of the above is an overall decrease in faith in the efficacy of multilateral treaties of arms control. This is of little direct implication for India since it was never a part of any of these, not even of the treaties that constitute the non-proliferation regime. However, the overall trend has indirect implications to the extent that New Delhi has always laid great score by multilateral arrangements as a measure of confidence building.

In the present environment, however, there appears to be greater propensity towards more bilateral or 'coalition of the willing' type of arrangements such as the Proliferation Security Initiative (PSI), Container Security Initiative (CSI)[55] or the United Nation Security Council Resolution (UNSCR) 1540[56]. While wary

55. The PSI and CSI are US led counter-proliferation initiatives that seek to conduct intelligence-based operations to detect or intercept cargo suspected of carrying WMD usable material. While India is yet to join the PSI, some of its ports are already CSI compliant.
56. The resolution was adopted in April 2004, soon after the revelations of the extent of the A.Q. Khan proliferation network. It mandated all countries to promulgate national legislation criminalising WMD proliferation and enforce effective export controls to minimise chances of WMD proliferation. National reports to that effect were to be submitted to a UNSCR 1540 Committee.

of such US led initiatives in the past, an economically resurgent India that has a new strategic relationship with the US, now finds it less threatening to accommodate these, if it considers them in its security interest.

Of course, there is no reason why India must stop pressing for more multilateral initiatives on arms control such as the negotiation of a universal, verifiable FMCT or the conclusion of a universal NFU treaty, and also on disarmament since a nuclear weapon free world (NWFW) is best suited to its security. Nevertheless, the country will have to be more patient and realistic on the chances of their acceptance. Meanwhile, India must also be prepared for eventualities such as the FMCT being exclusively negotiated by the five NWS through bargains that suit them and then put forth before the others for their signature. Given the US frustration with multilateralism, this could well be a possibility and India must weigh its options carefully before it becomes a reality.[57]

Blurring Lines Between Conventional and Nuclear Weapons

Until now, a clear distinction between conventional and nuclear weapons in terms of their doctrine, role, and delivery systems ensured a certain degree, however perverted, of stability. With the increase in lethality of conventional arsenals and the decrease in size and estimated damage of nuclear weapons, the lines could get hopelessly blurred. The impact of a weapon is determined on the basis of its power of destruction and as fuel air explosives (FAE), thermobaric bombs and electro-magnetic pulse (EMP) weapons show, they are not far behind nuclear weapons in destruction and impact.

In fact, advances in conventional weapons technology suggest that by 2020, precision long-range conventional weapons may be

57. This issue is explored in greater detail in Chapter 8 on Credible Nuclear Deterrence and Fissile Material Concerns.

capable of performing some of the missions currently assigned to nuclear weapons. For instance, uncertainty in the location of road mobile missiles carrying WMD requires a nuclear weapon for assured destruction today, but future real-time imagery and battle management, combined with precision strike long-range missiles, could use a conventional weapon to effectively destroy such targets tomorrow.[58] Hence, the US Administration has been interested in incorporating new conventional capabilities as being made available by the revolution in military affairs (RMA) into US strategic doctrine and force structure. In fact, the QDR 2006 had promised an initial capability to deliver precision-guided conventional warheads using long range Trident SLBMs within the next two years. In March 2008, the US Department of Defence did notify the Congress that it planned to conduct a flight test in 2009 on its Trident D-5 to determine how a modified, conventionally armed reentry vehicle might function on an SLBM that typically carries nuclear weapons.[59] Called Conventional Prompt Global Strike (CPGS), it is expected to provide the US with the capability to attack targets anywhere around the globe at a notice of 60 minutes. The Committee on CPGS concluded in 2008 that "there would be important political and strategic advantages to the US in being able to strike high value targets having time-sensitive urgency that could not be effectively engaged by currently available conventional strike systems."[60]

However, this trend could prove highly destabilising. In fact, several in the US are themselves unconvinced of its prudence,

---

58. For more on this, see Stephen M. Younger, "Nuclear Weapons in the 21st Century," Los Alamos National Laboratory document no. LAUR-00-2850, June 27, 2000.
59. "Navy to Flight Test Controversial Weapon Next Year," *Global Security Newswire*, April 3, 2008.
60. Naval Studies Board, *US Conventional Prompt Global Strike: Issues for 2008 and Beyond*, Report of 2008, available at http://www.nap.edu, p. 61

fearing particularly that Russia or China might mistake the launch of a conventional ballistic missile for a surprise nuclear attack requiring instant retaliation. The Congress, therefore, had rejected funding for such research in 2007 on the ground that the ambiguity created a risk. But the Pentagon pressed the Congress for $175 million to proceed with an initial phase of substituting conventional warheads for nuclear warheads on two SLBMs for each of the dozen deployed SSBNs, for targeting "fleeting, high-value, high-regret factor-type threats." [61]

The other problem arises from the fact that strategic arms reductions treaties do not cover strategic offensive weapons with non-nuclear warheads and it would be impossible to find out how many US strategic submarines have such warheads. For the time being, Russia has decried this move as "unfortunate".[62] But it might only be a matter of time before it and others follow suit, vastly complicating security scenarios and changing the nature of warfare. Any such move by China would have serious implications for India's security.

Increased Tussle Between Offence and Defence

The US has seen sagacity in complementing its deterrence capabilities with BMD in order to counter the current range of WMD threats. However, despite all the hyperbole over BMD, the fact remains that there can be no foolproof guarantee that every incoming missile shall be intercepted in time and every time. A certain amount of residual uncertainty would always remain. Meanwhile, military history purports that there is no defensive system that cannot be breached given sufficient time, technology, perseverance and offensive action by the adversary. In fact,

61. Wade Boese, "Global Strike Still on Pemtagon Wish List," *Arms Control Today*, April 2007.
62. Viktor Litovkin, "New American Non-Nuclear ICBM Creates Global Dangers," http://www.spacewar.com, June 16, 2006.

defensive capabilities often promote offensive behaviour. Russia and China are striving to reestablish strategic stability through vertical nuclear proliferation and the simultaneous development of BMD and counter-BMD technologies.

However, what also needs to be considered is that the US is factoring these measures into its BMD. For instance, in one R&D exercise, the Defence Advanced Research Projects Agency (DARPA) has conducted tests of a hypersonic projectile powered by a supersonic combustion ramjet (scramjet) engine burning hydrocarbon fuel,[63] as a step towards realisation of flight at hypersonic speeds, which could be used for powering long-range hypersonic missiles. Such developments indicate a race to seek superiority in offence and defence. The impact of this on inter-state relations is certain to be not all benign. Suspicious attitudes and actions premised on mistrust can naturally be expected to breed more of the same.

For India, the implications of US BMD come in an indirect manner in the form of instabilities generated by Russian and Chinese responses. Particularly, as China improves its offensive and defensive nuclear capabilities, it would have an impact on India's nuclear deterrence. While this aspect is studied in greater detail while examining Chinese nuclear thinking and strategy, it must however, be said that India will also be forced over time to sharpen both the offensive and defensive aspects of its nuclear arsenal.

Increased Insecurities and Proliferation

Given the trends highlighted above, it is hardly surprising that insecurities seem to be generally on the rise. At one level, the US-Russian relationship is going through turbulence unseen since the end of the Cold War. While the USA and the UK feel the need

63. "Hypersonic Scramjet Projectile Flies in Missile Test," http://www.spacedaily.com, September 4, 2001.

to keep a "weather eye open for a resurgent Russia,"[64] Moscow is wary of NATO's expansionist overtures and US BMD forays into its sphere of influence. In fact, Russia's discomfiture was clearly evident in a rather aggressive speech made by President Putin at the Munich Conference on Security Policy on February 10, 2007. While generally lambasting the US for its unilateral approach to issues and disregard for international law, Putin specifically attributed destabilisation of nuclear relations to two factors: one, the obvious stagnation in disarmament; and, second the plan to expand BMD into Europe.[65]

At another level too, the continued centrality of nuclear weapons in the security strategies of the NWS and the US propensity towards preemption has generated an extended impulse for the acquisition of nuclear weapons by countries that feel threatened by the US. According to an FAS report of 2006, eight countries launched more than 26 ballistic missiles of 23 types in 24 different events.[66] Along similar lines, a report by the WMD Commission put forth in 2006 estimated that ten new nuclear powers could emerge within the decade. Indeed, as long as nuclear weapons exist, non-proliferation will be a chimera. Contradictions between what the NWS are pursuing themselves and demanding of others is a dangerous game that will breed proliferation.

Not surprisingly, therefore, an assessment of the United Nations in the year 2005 was as pessimistic as, "The early 21st century is characterized by deteriorating relations among major powers, the unraveling of treaty-based security regimes, an increased sense of vulnerability to terrorist attack and crisis of legitimacy and credibility on national and international

64. Cobbold, n. 16, pp. 6-7.
65. Putin's Speech at the 43rd Munich Security Conference, February 10, 2007. As available at http://www.securityconference.de.
66. Martin Sieff, "8 Nations Now Building ICBM Nuclear Missiles," http://www.spacewar.com, December 26, 2006.

institutions of governance... Such developments are important factors in creating a fear filled security environment...."[67] Events since 2005, BMD deployments in Poland and Czechoslovakia, the war in Georgia in 2008, the ongoing concerns over the North Korean and Iranian nuclear ambitions, the state of Pakistan and risks from its nuclear arsenal, etc have only added to the complexities.

Such an environment then cannot be conducive to peace and economic development, the primary national security objectives of India. Since the country inhabits a region prone to proliferation (state to state, as well as non-state to state or vice versa, with or without government complicity) and terrorism, it faces real and urgent security challenges. While the country must address these through different means at different levels, the only sustainable way out of the nuclear action-reaction cycle is continued effort for a nuclear weapon free world and India must use every opportunity and forum to keep the issue alive, notwithstanding its dim prospects in the light of current trends in nuclear thinking and strategies.

Conclusion

Most NWS today concede that the role of nuclear weapons is very, very limited. However, the standard argument goes, "As long as other nations have stockpiles of nuclear weapons, we will need a robust nuclear deterrent to defend us and maintain a balance of power that stabilises the nuclear-armed world." Indeed, the world of nuclear weapons has existed as a globalised entity, in cause and effect, from much before the term acquired the modern day economic connotation. Any development in the nuclear arena of one country has normally had direct or indirect (through a cascading chain reaction) implications on another or more, which

67. Report of UN Secretary General, 2005.

has resonated with a proportionate or disproportionate response leading to more of the same.

As is evident from the examination of current nuclear developments, the end of the Cold War and the sweeping changes it brought about in international relations could not generate any significant change in the rigid mindsets that still envisage nuclear weapons at the centre of national security. New threats have meant additions and modifications to the concept of nuclear deterrence, not its removal from security strategy. While punishment remains the *modus vivendi* between NWS, BMD is expected to work against weaker adversaries from where the threat of retaliation is non- existent or manageable. The asymmetrical relationship, however, holds the temptation of preemption and, in turn, not only raises insecurities for others, but also increases the attractiveness of the nuclear weapons and breeds proliferation. If multilateral treaties are no longer able to hold countries to compliance with their commitments, the attraction of nuclear weapons could prove fatal.

Current trends in nuclear weapons thinking and strategies are hopelessly removed from the distinct possibility created in 1991 of actually discarding the weaponry as detritus of the Cold War. From the way the wind blows today, it is highly unlikely that another such opportunity will present itself any time soon. India's nuclear strategy must, therefore, take cognisance of this reality and gear up to meet the challenges of the future – by way of establishing a credible nuclear deterrent to deal with the realities of the day, while simultaneously pushing for more long-term security through the creation of a nuclear weapon free world.[68]

---

68. Many allege that there is a contradiction in this stance. Chapter 11 of this book discusses the subject in detail to explain the view of the author.

2

Shoot and Scoot
Pakistan's Nuclear Doctrine And Strategy

Ever since its independence, Pakistan has been central to India's national security consciousness. This assumed a more complex dimension with the introduction of nuclear weapons in the two nations. Even though the overt demonstration of Pakistan's nuclear weapon capability happened only on May 28 and 30, 1998, after India had revealed its nuclear hand, it is widely believed that Pakistan had acquired the necessary wherewithal in 1984 when A.Q. Khan claimed that the Khan Research Laboratories (KRL)[1] was "in a position to detonate... a nuclear device on a week's notice."[2] By 1986, the US National Intelligence Estimate had concluded that Pakistan was only "two

1. Inaugurated in July 1976, it was then known as Engineering Research Laboratories (ERL) until May 1981 when President Zia-ul Haq changed the name to Khan Research Laboratories in honour of A.Q. Khan's service to Pakistan's nuclear programme.
2. IISS Strategic Dossier, *Nuclear Black Markets: Pakistan, A.Q. Khan and the Rise of Proliferation Networks – A Net Assessment,* (London: International Institute for Strategic Studies, May 2, 2007), p. 22

screwdriver turns" away from a nuclear weapon[3], and the same was proudly acknowledged by Pakistan in 1987.[4]

Its nuclear capable status, even when not demonstrated through nuclear tests, gave Pakistan the confidence to follow a stratagem espoused by former Chief of the Army Staff, Gen Aslam Beg, designed to contain a conventionally superior Indian military through an offensive policy of engaging it in a proxy war with the help of *jehadi* groups. Over the last two decades, India has been trying to find the means of effectively dealing with the sub-conventional conflict cheekily indulged in by Islamabad from the shadows of its nuclear weapons.

To enable India to address this reality, it is important to delve into Pakistan's perception of its nuclear capability. This, of course, is closely linked to how the Pakistani military and political decision-makers (mostly the same) identify their country's security vulnerabilities, its major threat perceptions, and seek to redress them, including through the muscle flexing allowed by their possession of nuclear weapons. Pakistan holds its nuclear weapons as the *ultimate guarantor of national survival* and uses them intelligently to nullify India's conventional superiority. How does it do this? To what extent are nuclear weapons conceived as an extension of the country's conventional capability? What is Pakistan's threshold for the use of nuclear weapons – early in operations, following an escalatory spiral, or not at all? How does this posture reflect in its nuclear force structure and deployment strategy? What are the discerning features of its nuclear doctrine?

3. Ibid., p. 22.
4. Gen Aslam Beg, "Pakistan's Nuclear Propriety," *National Security*, FRIENDS, 2000, as cited in Paolo Cotta-Ramusino and Maurizio Martellini, *Nuclear Safety, Nuclear Stability and Nuclear Strategy in Pakistan*, Landau Network – Centro Volta Report, January 21, 2002. Gen Beg, Vice Chief of the Army in 1987 and Chief from 1988-91, claims that Pakistan had six nuclear devices by 1989 and 15 delivery systems by 1991.

These are some of the questions that the chapter considers in order to provide cues and directions to Indian policy-makers.

Pakistan's Expectations from its Nuclear Weapons

Pakistan considers the nuclear weapon as its most precious strategic asset. The country's leadership is extremely conscious of its value as the ultimate guarantor of the nation's existence as a sovereign entity. This was avidly brought out in one of the writings of Gen Mirza Aslam Beg. In an article appropriately titled "Pakistan's Nuclear Imperatives", he wrote, "Some safety against extinction is the inalienable right of an individual or a nation. Oxygen is basic to life, and one does not debate its desirability.... *nuclear deterrence has assumed that life-saving property for Pakistan.*"[5]

India is the *raison d'etre* of Pakistan's nuclear weapons. Its nuclear doctrine and strategy is *wholly and solely India-centric,* designed to address perceived conventional and nuclear threats from India. Consequently, the nature and function of the Pakistani nuclear deterrent (including delivery mechanisms), as also its rules of employment and deployment, are all tailored to meet this one requirement. And, from this one *brahmastra* in its repertoire of military capabilities, Pakistan, and particularly its army, given the nature of the system of governance in the country, expect to reap a rich harvest of military and political objectives.[6]

---

5. Gen Mirza Aslam Beg, "Pakistan's Nuclear Imperatives," *National Development and Security,* vol.3, no.10, November 1994, pp. 29-41. Emphasis added.
6. Peter Lavoy, in his essay on Pakistan's nuclear doctrine, has listed eight separate "uses" for Pakistan's nuclear weapons, four being specific to India, viz (i) last resort weapons to prevent military defeat or loss of territory; (ii) deterrent to conventional military attack; (iii) facilitators of low-intensity conflict; and (iv) tools to internationalise the Kashmir issue. See Rafiq Dossani, Henry S. Rowen, *Prospects for Peace in South Asia* (Stanford: Stanford University Press, 2005).

Military Objectives

Strategic Equaliser of Power Asymmetry

Pakistan has always resented its inherited geo-physical and structural asymmetry. A former Pakistani Foreign Minister, Abdul Sattar, lamented that the transition to independence "created seemingly impossible problems for Pakistan, which unlike India, inherited neither a capital nor government nor the financial resources to establish and equip the administrative, economic and military institutions of the new state...."[7] This perception that "Pakistan started its independent career as a weak nation,"[8] for which India was held blameworthy, was met by defining "national identity through religious symbolism and by building India-Pakistan rivalry."[9] Ever since, Pakistan has looked for ways and means to somehow equalise the power asymmetry with India through alliance building with the USA, China and other Muslim countries, as well as through acquisition of modern conventional weaponry to match a far larger and better equipped Indian Army. However, it was in the acquisition of a nuclear weapon capability that Pakistan discovered the best and most effective equaliser.

Islamabad had begun to consider acquisition of nuclear weapons from the time China tested its own nuclear device in 1964 because it assumed that this development would set India down the nuclear path. Zulfikar Ali Bhutto, Pakistan's Foreign Minister in the government of Gen Ayub Khan, emphasised the deterrence

7. Abdul Sattar, "Fifty Years of the Kashmir Dispute: The Diplomatic Aspect," in Suroosh Irfani, ed., *Fifty Years of the Kashmir Dispute* (Muzaffarabad: University of Azad J & K, 1997).
8. Pervaiz Iqbal Cheema, *The Armed Forces of Pakistan* (Karachi: Oxford University Press, 2003), p. 34.
9. Hussain Haqqani, *Pakistan: Between Mosque and Military* (Lahore: Vanguard Books, 2005), p. 12. The author argues that focus on rivalry with India was used as an instrument by the national security apparatus in Pakistan to secure legitimacy for themselves. This continues to be the case.

value of nuclear weapons in 1965 and pressed upon his president that, "All wars of our age have become total wars... and our plan should, therefore, include the nuclear deterrent."[10] However, it was only in January 1972, within three weeks of its defeat in the 1971 War that serious thought was given to the nuclear weapon as an effective instrument to match up to a larger power.[11] Pakistan was not unique in reaching this conclusion. Countries with inferior conventional capabilities seeking to deter larger states with superior militaries have found the nuclear option a useful one. Even the British Prime Minister, Margaret Thatcher had said in Moscow on March 31, 1987, "Nuclear deterrence is the only means allowing small countries... to stand up to big countries."[12] Likewise, in its search for viable security, Pakistan has found in nuclear weapons a means to balance India's conventional superiority.

At the same time, the nuclear weapons programme has provided Islamabad the additional benefit of addressing India's perceived advantage in science and technology. In this context, it merits remembering that Z. A. Bhutto, while he himself was out of office in 1969, had cautioned the then Pakistani government, "If Pakistan restricts or suspends her nuclear programme it would not only enable India to blackmail Pakistan with her nuclear advantage, but would *impose a crippling limitation on the development of Pakistan's science and technology.*"[13] He was clear that Pakistan's development of nuclear weapons would

---

10. Z.A. Bhutto, *The Myth of Independence* (Karachi: Oxford University Press, 1969), p. 153.
11. Jasjit Singh has identified two crucial strands of the Pakistani nuclear strategy that evolved after 1971: (a) nuclear weapons to neutralise India's conventional and military superiority; (b) use of irregular forces and terrorism. Jasjit Singh, "Pakistan's Nuclear Strategy: An Assessment," *Aakros*h, vol 4, no.13, October 2001, p. 73.
12. Gerard J. Degroot, *The Bomb: A History of Hell on Earth* (London: Pimlico, 2004), p. 188.
13. Bhutto, n.10. Emphasis added.

serve to effectively and more holistically counter-balance India's preeminence in the subcontinent, and demonstrate national technological capabilities.

In yet another use of its nuclear weapon for bridging the gap with India, Pakistan has intelligently manipulated the US for conventional arms and economic assistance. For instance, after India's peaceful nuclear explosion in 1974, Bhutto warned the US that if his country was unable to get "sufficient conventional weapons" to bridge the disparity, Pakistan "would forego spending on conventional weapons and make a big jump forward for concentrating all its energy on acquiring the nuclear capability."[14] Pakistani leaders have always tried to make a convincing case to justify nuclear weapons as the only means available to "preserve a broad strategic equilibrium with India, to neutralise Indian nuclear threats or blackmail, and to counter India's large conventional forces."[15] And, the US was easily convinced since it anyway perceived India as part of the Soviet constellation in the Cold War, while Pakistan was their ally against the Communist bloc.

Deterrent Against Conventional War

Common sense mandates that nuclear deterrence premised on mutual assured destruction should place an automatic limitation on violence and act as a brake on total war. Given its horrendous consequences, no rational leader could risk war if nuclear retaliation was even remotely possible. This reality has been intelligently exploited by Pakistan in its national security strategy. Islamabad believes that notwithstanding its conventional

14. P. B. Sinha & R.R. Subramanian, *Nuclear Pakistan: Atomic Threat to South Asia* (New Delhi: Vision Books Pvt Ltd., 1980), p.70.
15. Zafar Iqbal Cheema, "Pakistan's Nuclear Use Doctrine and Command and Control", in Peter Lavoy, Scott D. Sagan and Jim Wirtz, eds., *Planning the Unthinkable: How the New Powers Will Use Nuclear Biological and Chemical Weapons* (Ithaca: Cornell University Press, 2000), p. 169.

inferiority, the nuclear weapon provides it an "infrangible guarantee of its independence and physical integrity"[16] *even* as it pursues provocative and offensive revisionist policies. Keen to avoid a full-scale conventional conflict with India, but desirous of altering the status of Kashmir through a constant needling of India, Pakistan reckons this is possible through nuclear weapons. Without them in its quiver, it is certain that Pakistan's proxy war would have breached the limits of India's tolerance and led to a conventional offensive. This eventuality, Pakistan perceives, is blocked by its nuclear weapons. As explained by Benazir Bhutto in an interview in 2004, nuclear weapons ensured that "India could not launch a conventional war, knowing that if it did, it would turn nuclear, and that hundreds of millions would die... suicide not just for one, but for both nations."[17]

Evidently then, Pakistani leaders, even civilian, have believed that the danger of nuclear escalation insulates Pakistan from an Indian conventional attack, allowing Pakistan to not only ensure its own security, but to also blatantly pursue an aggressive strategy in Kashmir. Pakistan's willingness to launch the Kargil conflict was based on the hypothesis that "their new overt nuclear status would enable them to deter the Indians even more effectively than their *de facto* nuclear capacity had previously done."[18] Two other assumptions underlay the adventure: profound confidence in the belief that the political establishment in India, especially the caretaker government then in power, would not be able to take a hard decision that could carry a risk of escalation; and, a reasonable

16. Ian Smart, The Great Engines: The Rise and Decline of a Nuclear Age", *International Affairs*, vol.51, 1975, p. 548 as cited by Rifaat Hussain "Nuclear Doctrines in South Asia", SASSU Research Report no.4, December 2005.
17. S. Paul Kapur, "India and Pakistan's Unstable Peace: Why Nuclear South Asia is Not Like Cold War Europe," *International Security*, vol.30, no.2 Fall 2005, pp. 127-152.
18. Ibid., p. 144.

confidence that the presence of nuclear weapons would attract the immediate attention of third parties (most notably, the USA) anxious to defuse a potential nuclear confrontation in South Asia. These conditions were expected to force India to accept a stalemate even at the cost of losing a small slice of its territory. However, that all the three assumptions proved untrue is another story. Yet, for the Pakistani security establishment, nuclear weapons continue to provide a shield to deter conventional aggression, even as it pursues covert aggression through sub-conventional means.

Facilitators of Sub-Conventional Warfare

In most countries possessing nuclear weapons, these are mainly considered instruments of nuclear deterrence, meant to obviate a *nuclear* attack by the enemy. But for Pakistan, its nuclear weapons serve to not only deter a *nuclear* and a *conventional* attack, but also provide it with the immunity to indulge in aggressive military strategies that harbour political ambitions. Accordingly, the Pakistani nuclear doctrine encapsulates a more offensive form of deterrence that seeks to change the status quo by holding out the threat of nuclear blackmail on Kashmir while deterring an Indian conventional attack.

This linkage was evident once there was a significant spurt in insurgency and terrorism in Jammu and Kashmir (J&K) from 1988 onwards, soon after the acquisition of nuclear weapons capability by Pakistan. The trend continued through the 1990s and was interspersed by statements by A.Q. Khan in 1992, and other political and military leaders in 1995-96 threatening India from taking decisive military action against Pakistan.[19] The idea was to allow Islamabad the freedom to manage escalation at a desirable level. According to one analyst, "Islamabad is convinced

19. For more on these linkages, see J.N. Dixit, *India-Pakistan in War and Peace* (New Delhi: Books Today, 2002).

that the mere threat of approaching the nuclear threshold will prevent India from seizing the strategic initiative and military dominance of events, permitting Pakistan to escalate the crisis at will without the fear of meaningful Indian retribution."[20] It was hardly surprising, therefore, that even amidst fighting in Kargil, Pakistan Army leaders insisted, "There is no chance of the Kargil conflict leading to a full-fledged war between the two sides."[21] Interestingly, this was similar to the advice given by senior US military officers to President Kennedy during the Cuban crisis in 1962. They believed that the US could afford to launch a limited attack on Cuba because the USSR *would not dare* counter-attack in Berlin. Pakistani military counsel to the civilian government too dismissed the possibility of a total war because nuclear deterrence afforded it greater impunity and immunity.

This is also the message Pakistan imbibed from India's decision not to launch an invasion in response to a provocative terrorist attack on the Indian Parliament in 2001 and the Kaluchak massacre at the height of troop mobilisation during Operation Parakram. Buoyed by the perception that nuclear deterrence worked in their favour, the larger Pakistani strategy appears to be to continue to bleed India while not provoking it enough to escalate to a point where any kind of decisive Indian action might threaten Pakistan.

India, meanwhile, has refuted this presumption by repeatedly alluding to the space for a limited conventional conflict below the nuclear threshold. In fact, in the context of the US-USSR nuclear equation, Glenn Snyder had extrapolated in 1961, "The Soviets probably feel... that there is a range of minor ventures which they can undertake with impunity, despite the objective existence of

20. Yossef Bodansky, "Pakistan's Nuclear Brinkmanship," Freeman Centre for Strategic Studies, Israel. Available at http://www.freeman.org
21. Ihtasham ul Haque, "Peace Linked to Kashmir Solution," *Dawn Weekly Wire Service*, June 26, 1999. As cited in Lavoy, n.15, pp. 248-249.

some probability of retaliation."[22] India believes the same. The danger, however, remains that neither side is, or can be, absolutely clear about where the threshold lies. And, this, or the fear that conflicts can take on a dynamics of their own which makes escalation difficult to predict or control, imposes limitations on the initiator of violence. Therefore, even as Pakistan continues to use its nuclear weapons as effective facilitators of sub-conventional war to complicate India's security challenges, it cannot, at the same time, be free of the fear of an unintended escalation.

Nuclear Deterrence

Amongst their most prosaic military functions, nuclear weapons in Pakistan perform the traditional task of deterring the adversary's use of nuclear weapons. MAD is expected to uphold deterrence as long as both sides have a retaliatory capability and neither has a first strike capability. So it is that Pakistan, in order to enhance the credibility of its deterrence, has been working towards acquiring a second strike capability against India. Its concept of nuclear deterrence, however, is built on projecting the nuclear weapon as a militarily usable one. By indicating its willingness to use the weapon for its survival, it seeks to enforce deterrence against India's nuclear and conventional forces.

Political Objectives

Security Guarantor in Absence of Alliance Support for Military Adventurism

Islamabad embarked upon alliance building within the first decade of its independent existence to buttress its security—a Mutual Defence Alliance with the USA in 1953, further reinforced through

22. Glenn Snyder, *Deterrence and Defense* (Princeton, NJ: Princeton University Press, 1961), p. 226.

an executive arrangement in 1959. This was seriously pushed for conversion, in the late 1970s, into a formal security treaty that would strengthen US security guarantees.[23] Though this was not formalised, President Nixon had already declared in September 1973, "Independence and integrity of Pakistan is a cornerstone of American foreign policy." The US became its major power patron, showering weapons and economic aid on Islamabad. Meanwhile, after eleven long years of negotiations, Pakistan finally established a close relationship with China through a bilateral agreement in June 1976. The historic import of this was described by Bhutto himself as "my greatest achievement and contribution to the survival of our people and our nation."[24]

While these relationships translated into a rich haul of military hardware and other assistance (including nuclear materials such as ring magnets, special furnaces, heavy water, tritium, weapon designs and even the conduct of a nuclear test by China), Pakistan never found them very forthcoming with support for its territorial skirmishes with India. The US disappointed Islamabad in 1965 and 1971, and it was equally miffed with China's advice to withdraw its forces from Indian territory in 1965. In 1999 too, both were loath to provide material or moral support to Islamabad for its adventure in Kargil. Therefore, over a period of time, and after every war with India, Pakistan has reluctantly accepted that neither the USA nor China would opt to become embroiled militarily in Indo-Pak Wars, and, in fact, would not hesitate to impose sanctions and embargoes on it. This made Islamabad realise its own vulnerability to political manoeuvring as a result of its dependence on external sources of

23. For more on this, see Savita Datt, *To Chagai and Beyond: Nuclear Developments in Pakistan* (New Delhi: IK International Pvt Ltd, 2003), pp. 42-44.
24. Zulfikar Ali Bhutto, *If I am Assassinated...* (New Delhi: Vikas, 1979) p. 138. This has been interpreted as an expression of Bhutto's satisfaction at having entered into an agreement with China that envisaged assistance to Pakistan's nuclear programme.

weaponry. These realisations have added to the Pakistani perception that only its own nuclear weapons capability could guarantee its security and survival.

This consideration weighed heavily upon the Pakistani administration in 1998 when after the Indian nuclear tests, Washington strongly advised Islamabad to refrain from a similar exercise and even promised substantial economic and military assistance, including the lifting of the Pressler Amendment[25], delivery of the 70 F-16s that had been stalled earlier, and other new weaponry. However, what was not forthcoming was the assurance of continuous military support against India. And, as far as Pakistan was concerned, that was reason enough to overtly demonstrate its nuclear capability. Two other assumptions helped it to make the decision: one, assurance provided by China to help Pakistan in the face of sanctions that were sure to follow its nuclear tests[26]; and, second the confidence that by showcasing the region as a dangerous nuclear flashpoint, it would anyway be able to extract concessions from the US. In fact, Kargil, it was assumed, would bring international attention to the region and Pakistan had certainly not bargained for the negative vibes that it received from its allies. The Kargil experience, therefore, reinforced Pakistani faith in its nuclear deterrent as the most reliable tool for national security. Meanwhile, as India transforms its relations with the US and China, the Pakistani fear psychosis and its reliance on nuclear weapons would only grow.[27]

---

25. The US Congress passed this legislation in 1985 requiring the US president to certify that Pakistan did not possess a nuclear weapon as a precondition for American economic and military assistance.
26. Pakistani officials did travel to China in the period after India's and before its own nuclear tests.
27. Interestingly, some Pakistani analysts have attributed the Pakistani desire for nuclear weapons to Sino-Indian rapprochement since that would erode Pakistan's margin for diplomatic manoeuvring.

Tool for Nation Building and Prestige

Apart from security, Bhutto's push for nuclear weapons was also motivated by a desire to "divert the nation's attention from the humiliation it had suffered as a result of its defeat in the East..."[28] and his yearning for Pakistan "to walk tall".[29] Even President Musharraf has referred to Pakistan's nuclear achievements in the same vein. In a speech delivered on March 27, 2001 on the occasion of the retirement of A. Q. Khan, he said, "In a general sea of disappointment, the development of Pakistan's nuclear capability is a unique national success story."[30] And the nation and its people second this opinion. The country's nuclear achievements are an effective rallying point for national pride. Not surprisingly, therefore, Kahuta that houses the uranium enrichment plant has been described as the "symbol of our national defiance" and the mastery of the enrichment technology as the "symbol of national pride, scientific and technical modernity and independence from foreign powers."[31] This has been explained thus by Brahma Chellaney[32]:

> The rapid technological advances by Pakistan in recent years are a symbol of nationalistic pride in a country which has overcome major political, technical, and industrial challenges to mount a program with a team of dedicated scientists. Pakistan is showing the world—as China did in the sixties--how a country with limited technical resources and a narrow industrial base can acquire nuclear weapons and ballistic missile capabilities by riding a wave of nationalism.

---

28. Datt, n.23.
29. Steve Weissman and Herbert Krosney, *The Islamic Bomb* (New York: Times Books, 1981), pp. 17-18.
30. Gordon Corera, *Shopping for Bombs: Nuclear Proliferation, Global Insecurity and the Rise and Fall of the AQ Khan Network* (New Delhi: Foundation Books, 2006), p. 150.
31. IISS Strategic Dossier, n.2, p. 20.
32. Brahma Chellaney, "South Asia's Passage to Nuclear Power," *International Security* vol. 16, issue 1, 1991, p. 43.

Instrument to Legitimise Military Power

Ironically enough, Zulfikar Ali Bhutto originally intended Pakistan's nuclear programme to serve as a civilian counterweight to the military. The programme was launched by his civilian government and the military evinced little interest in the same. In fact, Gen Ayub Khan as president, had dismissed Bhutto's calls for attention to nuclear weapons when he was his foreign minister from 1963-66. But, it was during Zia-ul Haq's tenure as president, that the programme came under military control and direction. And there it has remained ever since. In fact, the Pakistan Army, keeping even the Air Force and Navy out[33], has effectively honed it as a "trump card"[34] against the civilian politicians and not allowed their influence in nuclear decision-making. Benazir Bhutto has openly admitted her limited participation in the country's nuclear decisions, including that she was allowed to be the prime minister under the tacit agreement that she would not interfere with the military's control of the nuclear programme. Meanwhile, by constantly retaining the domestic focus on rivalry with India, the Pakistan Army has conditioned the nation's public to believe that there is a constant threat from India, which can only be effectively addressed by allowing the military a free hand with the country's economy, polity and nuclear policy.

Tool for Gaining Leadership of the Islamic World

Besides using it as a rallying point for national pride, Pakistan has also used the concept of "Islamic Bomb" to acquire support for, and leadership of, the Islamic countries. By portraying its nuclear weapons capability as belonging to the larger Muslim community,

---

33. The army has monopolised the post of Chairman, Joint Chiefs of Staff since the 1998 nuclear tests, abandoning the tradition of all the three Services holding the position by rotation.
34. Ashley Tellis, *India's Emerging Nuclear Posture: Between Recessed Deterrence and Ready Arsenal* (Santa Monica: RAND, 2001).

especially in the earlier decades of its nuclear programme, Pakistan was able to garner material and moral support from other Muslim countries such as Saudi Arabia, Libya and Iran. President Zia said in 1978 that Pakistan's possession of the nuclear weapon "would reinforce the power of the Muslim world."[35] Twenty years later, after the nuclear tests in Chagai, Prime Minister Sharif said, "Not only the whole nation, but the *whole Islamic ummah* hailed Pakistan for its great achievement and expressed happiness over the decision."[36]

However, this tendency to project a wider belongingness of the Pakistani nuclear weapon has considerably reduced post 9/11 and after the revelations of the A.Q. Khan network. Islamabad is now more conscious of signalling a firmer control and more responsible attitude towards its nuclear assets. This is not to suggest that earlier Islamabad had allowed other nations any physical access to its nuclear arsenal, but it did encourage a notional sense of wider Islamic pride in its nuclear venture, which in some cases as Iran and Libya translated into nuclear proliferation. It is, however, now better known that Pakistan was clever enough to export only snippets of information and obsolete technologies and equipment. This may have been a clever ploy to retain its status as the only Muslim country possessing nuclear weapons.

Pakistan's Nuclear Doctrine

There is no official Pakistani nuclear doctrine. Whether as a matter of deliberate policy, or purely by default, but probably because of a bit of both, Pakistan's nuclear doctrine is shrouded in ambiguity. This is not surprising given that most states possessing nuclear weapons seek to exploit the role of opacity

35. Smruti Pattanaik, "Pakistan's Nuclear Strategy," *Strategic Analysis*, vol 27, no.1, January-March 2003.
36. *Summary of World Broadcast*, Part 3 (Asia Pacific), FE/3241, June 1, 1998, p. A1 . Emphasis added.

and ambiguity by refusing to define the number of weapons in their arsenal or the precise trigger points for nuclear use. So it was that a former Pakistani Foreign Minister, Agha Shahi wrote in 2000:

> ...a policy of ambiguity would appear to be best for Pakistan's security. Spelling out its nuclear doctrine would detract from the imperative of uncertainty about when a nuclear strike as a last resort would... reinforce maximally credible nuclear deterrence by raising the threshold of Indian calculation of unacceptable risk.[37]

Despite the lack of transparency, however, it is possible to draw out some contours of the Pakistani nuclear doctrine from the statements of civilian and military leaders and from the writings on the subject by the Pakistani strategic community.

Minimum Nuclear Deterrence

Like the doctrines of India and China, Pakistani official statements too advocate minimum nuclear deterrence. One year after the conduct of its tests, in May 1999, Prime Minister Nawaz Sharif, in a lecture to the officers of Pakistan's National Defence College (NDC) described minimum nuclear deterrence as "one of the principles guiding Pakistan's nuclear policy."[38] Foreign Minister Abdul Sattar reiterated the same while addressing an international seminar in Islamabad on November 25, 1999. It has been articulated that Pakistani nuclear deterrence *does not* stem from the quantity of its weapons, but from "its quality, which is primarily a function of the Pakistani leadership to pursue a 'no holds barred' approach towards defensive use of nuclear weapons in the event of a war

37. Agha Shahi, "Command and Control of Nuclear Weapons in South Asia," *Strategic Issues*, March 2000, p.56, as cited in Hussain, n. 16, p. 13.
38. Rajesh Rajagopalan, *Second Strike: Arguments About Nuclear War in South Asia* (New Delhi: Penguin, 2005), p. 51.

with India."[39] Therefore, Islamabad, like New Delhi, maintains that the "minimum" cannot be defined and is, in fact, a dynamic concept based on calculations of the number of nuclear weapons estimated with the adversary, the manner of their deployment, and other technological imperatives such as missile defence. Since these factors impinge upon the survivability of nuclear weapons, and, hence, the deterrence credibility, the quantity of weapons would require to be "upgraded in proportion to the heightened threat of preemption and interception."[40]

Therefore, Pakistan's calculation of minimum is closely linked to its perception of India's nuclear numbers, and its ability to inflict unacceptable damage on India. While the first parameter may be interpreted to mean parity, the second conveys that for nuclear deterrence to work, parity need not be based on "numerical equality of the number of nuclear delivery systems, or of the number of warheads or in the yield of megatons available to each opponent. Parity requires assured destruction capability."[41] The problem, however, arises in defining what would constitute unacceptable damage for a geographically bigger, and an institutionally and economically stronger India. Gen Mahmud Durrani (Retd) sought to resolve this problem by suggesting, "Overkill would by necessity be built into the response."[42]

Pakistan's construction of minimum nuclear deterrence, however, is not very different from that of India. It is a concept that is less definable in terms of numbers and more a reflection of the relatively small amount of resources – financial as well as

39. Ibid., p. 14.
40. Ibid., p. 49.
41. Zafar Nawaz Jaspal, "Reassessing Pakistan's Nuclear Strategy," http://www.defencejournal.com, July 2001.
42. Maj Gen Mahmud Durrani, "Pakistan's Strategic Thinking and Role of Nuclear Weapons," *Cooperative Monitoring Centre Occasional Paper*, SAND, 2004, available at http://www.cmc.sandia.goy.

fissile – being put to optimum use for enhancing the credibility of deterrence.

Nuclear First Use

Pakistan's main concern has always been to offset India's superior conventional military and its own lack of strategic depth that constraints its ability to pursue a defensive strategy. The nuclear doctrine is aimed at addressing these limitations and, hence, it is not at all illogical that Islamabad retains the option of first use of nuclear weapons. Fearful of the possibility of finding itself at the receiving end in a conventional conflict[43], Pakistan projects its nuclear arsenal as a weapon of last resort. As argued by Air Cmde Tariq Mahmud Ashraf (Retd):

> Being on a weaker military footing... Pakistan's nuclear employment doctrine should assert that since she would be fighting for her very survival as an independent nation state in any future war,... Pakistan... must reserve the right of first use of nuclear weapons and this assertion should be made as part of her nuclear employment doctrine.[44]

NATO, it may be recalled, followed the same policy in the face of Soviet conventional superiority through the Cold War years. Along similar lines, viewing 'first use' as an "option enhancing policy,"[45] Pakistan dismisses India's offers for a mutual no first use agreement. As explained by Lt Gen Sardar Lodi (Retd),

---

43. Pakistani strategists point out that India's capture of just 140 km would wipe out the state because its communication, irrigation, industry and population are all together within that depth. See Syed Rifaat Hussain, "War Against Terrorism: Pakistan's Perspective," *IPRI Journal*, vol 4, no.1, 2004.
44. As cited in Hussain, n. 16, p. 11.
45. Stephen Cohen, *The Pakistan Army*(London: Oxford University Press, 1998).

> India's offer of a treaty to be signed by the two countries, agreeing not to be the first to use nuclear weapons against each other is one-sided and would benefit India only, as it has a superior conventional force. It may be more apt for both countries to sign a mutual test ban treaty to start with, followed by a no-war pact.[46]

However, for all its emphasis on first use, India need not interpret first use as early use of nuclear weapons in a conflict. Rather, this posture of a low nuclear threshold stems from a national compulsion because Pakistan's offensive strategy at the sub-conventional level keeps it in a state of perpetual conflict.[47] This is meant to reinforce its deterrence since nuclear weapons become an effective shield even against conventional attack. The need for Pakistan to project such a position is further exacerbated when India insists that there is the possibility of a limited conventional war even in a nuclear environment. This formulation troubles Pakistan because it interferes with its basic rationale for acquisition of nuclear weapons, which is to deter India from using its conventionally superior force against them. Therefore, in order to stave off such an eventuality, it seeks to deter India with the thought that Pakistan would not hesitate to use nuclear weapons if pushed against the wall.

But, how far back would be too far back has obviously not been clearly identified. In 1999, an op-ed by three prominent Pakistani leaders stated that Pakistan would resort to a nuclear strike only in the event of comprehensive military defeat, threat to large

46. Sardar Lodi, "Pakistan's Nuclear Doctrine," *Defence Journal*, vol.3, no.4, April 1999.
47. Ajay Behera, "On the Edge of Metamorphosis" in Ajay Behera and Joseph C. Mathew, eds, *Pakistan in a Changing Strategic Context* (New Delhi: Knowledge World, 2004), p. 36.

population centres, or lines of communication.[48] Subsequently in 2002, Gen Khalid Kidwai, head of Strategic Plans Division (SPD) that manages Pakistan's nuclear assets and operations, spelt out four distinct thresholds for nuclear use: loss of large parts of territory (space threshold); destruction of large part of land or air forces (military threshold); economic strangulation (economic threshold); and political destabilisation or large scale internal subversion (domestic destabilisation threshold). Since then, however, Pakistan has been quieter on its red lines. In fact, Gen Asad Durrani even said in 2003, Pakistan does not "identify those core interests that, if threatened, could trigger a nuclear retort... These are elements of operational planning and stating them could betray a country's conventional limits." [49]

By vaguely articulating a range of thresholds likely to trigger nuclear first use, Pakistan has sought to strengthen its strategy of continuing sub-conventional conflict while checkmating India's superior conventional capability. However, it has been suggested by some Pakistani military officials that it is most likely that instead of escalating a conflict itself, Pakistan will leave escalation to India, though it would not hesitate from provoking it to a point where Islamabad gets reason to go nuclear! Lt Gen Javed Hassan, commandant of the NDC of Pakistan argued along these lines at the Brookings Institution in 2002.[50] He opined that if India applied a great deal of pressure in one sector across the Line of Control (LoC) or the international border (IB), Pakistan would respond with disproportionate counter-force which would compel India to

48. Agha Shahi, Zulfiqar Ali Khan and Abdul Sattar, "Securing Nuclear Peace", *The News* (Islamabad), October 5, 1999.
49. As cited in Varun Sahni, "The Stability-Instability Paradox: A Less than Perfect Explanation" in E. Sridharan, ed., *The India-Pakistan Nuclear Relationship: Theories of Deterrence and International Relations* (New Delhi: Routledge, 2007), pp. 185-207.
50. As cited in Behera, n. 47, p. 36.

escalate beyond a sector to engage Pakistan all along the LoC or the IB and if India's escalation crossed the Pakistani nuclear threshold, it will have cause and justification to escalate to the nuclear level. "India will have been shown to have behaved irresponsibly and forced Pakistan to take extreme measures."

Such calculations, however, do not stand the test of rationality for they fail to consider the consequences of the same for Pakistan. Once Pakistan has used nuclear weapons against India, New Delhi could respond in only two ways to the act: retaliate with, or without nuclear weapons. Pakistan assumes that even after being hit with a nuclear weapon, India would not necessarily undertake nuclear retaliation for three reasons: one, given the Indian strategic culture, New Delhi would not have the stomach and the will to undertake nuclear retaliation; second, the major powers would intervene to stop India, especially if Pakistan had undertaken a small first strike against counter-force targets; third, with Pakistan's slow but steady acquisition of a second strike capability, India would be deterred from nuclear use for fear of further nuclear strikes. One or a combination of these circumstances, it is assumed, would not only allow Pakistan to get away with its nuclear use, but also enable it to achieve its political objectives.

However, these are only assumptions and there is nothing to indicate that they might not prove false during the moment of truth. And if, contrary to Pakistani expectations, India did respond with its nuclear arsenal in a decisive fashion to put an end to the "bleeding through a hundred cuts policy," then the consequences for Pakistan could be fatal. In fact, unlike Pakistan, Indian writings have sought to describe not what India might do in the event of deterrence breakdown, but rather what needs to be done to prevent such a breakdown from occurring and this perhaps is responsible for the mistaken impression that India might not retaliate even after a nuclear attack. It may be stated that though India has shown itself chary of using the military option against Pakistan-fomented terrorism, the use of nuclear weapons would certainly galvanise the country into action for the simple reason that it cannot allow any erosion of credibility of its deterrence for the future. Unlike Pakistan that would be severely set back by India's nuclear use, India would survive, and for the sake of India's future, a retaliatory nuclear response would become mandatory. Pakistan must, therefore, rethink its calculations in this regard.

Use Against Conventional Force

Just as the US nuclear doctrine maintained a constant underlying theme that nuclear weapons would be employed against the USSR to offset its numerical superiority in manpower and conventional arms, Pakistan too has the same doctrine. As said by Sardar Lodi,[51]

> In a deteriorating military situation when an Indian conventional attack is likely to break through our defences or has already breached the main defence line causing a major setback to the defences, which cannot be restored by conventional means at our disposal, the government would be left with no other option except to use nuclear weapons to stabilize the situation.

Therefore, unlike India where the nuclear weapon is perceived as a special and distinct weapon of immense destructive potential that defies rational use, except for enforcing deterrence, Pakistan seeks to systematically integrate it as another weapon into its operational military planning. For India, the impact of the weapon is politico-psychological, while for Pakistan it is military-operational.[52] Islamabad treats its nuclear arsenal as an integral

---

51. Lodi, n. 46.
52. Tellis, n. 34, p. 301.

element of its crisis management and military strategy. In fact, analysts like Shireen Mazari have suggested that Pakistan should project its nuclear use as being based on a "one rung escalation ladder *knitted tightly with a highly cohesive, state-of-the-art tactical conventional military*."[53]

Offensive Defence

Given the above, it is natural that Pakistan's nuclear doctrine should be no different from its overall military doctrine of offensive defence. The nuclear weapon is projected both as an offensive instrument that holds out the threat of nuclear strike right at the beginning of every crisis, irrespective of its nature and scale, and, at the same time, also a defensive weapon against India's punitive action. This is meant to give Pakistan an opportunity to conduct a swift, sub-conventional assault, mostly in Kashmir, and then use the nuclear shield to prevent/mellow/checkmate an Indian response.

Therefore, there is a huge doctrinal gap between India and Pakistan on this issue. While India sees only retributive utility in nuclear weapons, Pakistan seeks to derive deterrence by propagating the nuclear weapon for "total defence", to deter both nuclear and conventional attack, and against counter-force and counter-value targets. Lt Gen Kamal Matinuddin (Retd) had written in 2002 that "[Indian] population centers, industrial assets, resources, and nuclear or conventional forces of the enemy can all be targeted... While giving primacy to counter-value targets, the enemy's concentration of armoured formations in the rear should also be considered as targets for a nuclear strike."[54]

53. Unpublished paper on South Asian Peace and Security Doctrine, presented at a seminar in Dhaka in 2000 and as cited by Pattanaik, n.35. Emphasis added.
54. As cited in Hussain, n. 16, p. 13.

The irony in Pakistan's nuclear doctrine is that it seeks security in the idea that nuclear war cannot be prosecuted for any rational political objective, while attempting the same through its own weapons.

Requirements of Pakistan's Nuclear Deterrence

To checkmate a possible conventional attack from a superior Indian force, Pakistan's nuclear deterrence needs to be sufficiently credible. Its capability, force posture and force structure must convey the message that it is capable of using the nuclear weapon first and yet somehow deny the choice of retaliation to India. In order to perform this task credibly, the Pakistani nuclear force must essentially meet the following requirements.

Not be Too Small a Nuclear Force

Despite professing minimum nuclear deterrence, Pakistan is cautious not to build too small a nuclear force since "small forces would presumably be easier to destroy in a first strike and, therefore, have less credibility as a deterrent because the surviving forces may not be able to retaliate."[55] In fact, though immediately after the tests, Pakistan was satisfied with existential deterrence based on uncertainty, it now believes that development of a second strike capability is critical for credible deterrence. In this, it may have been influenced by writers such as Andre Beaufre who had suggested in the case of France that "minor nuclear powers can deter much larger nuclear forces if they can hold out the threat of nuclear riposte, even if the riposte is weak."[56]

This desire could expectedly grow as India acquires and deploys some form of a limited ballistic missile defence (BMD). Pakistan strongly believes that such a capability with India, coupled with

55. Jaspal, n. 41
56. Rajagopalan, n.38, p. 27.

Pakistan's lack of strategic depth would destabilise its deterrence. Islamabad fears that such a capability would "provide an elated sense of security and prompt preemptive impulses from India."[57]

Pakistan's response to this is likely to be to go in for its own BMD systems as well as building up its offensive forces to overwhelm Indian defences. It is not a stray coincidence that the rate of nuclear and missile activity in Pakistan has risen manifold over the last few years. Former President Musharraf admitted as much in July 2005 when he said the country's nuclear programme is progressing "ten times faster than before."[58] While there might be an element of exaggeration here, undoubtedly, the pace of the programme has accelerated. And for buttressing its nuclear defensive and offensive capabilities, it is likely to benefit from its all weather friend, China which has done substantial research and development on BMD and counter-measures.

Capable of Quick Assembly

It has been stated by Pakistani officials, including Gen Musharraf, that Pakistani nuclear weapons exist in a "disassembled state". However, the level of disassembly, and naturally, therefore, the time taken for assembly, is not clear. It could mean a state where the weapons i.e. the frame and fissile core, are kept separately at a storage facility, and delivery systems are kept elsewhere; or where weapons (frame and core) and delivery systems are stored at the same military base for rapid mating; or, where weapons (fissile cores) are kept separately but the frame is already mounted on delivery systems. The Pakistani nuclear arsenal is presently believed to be in the third state of disassembly.

57. Fear, as stated by no less than Director, Arms Control and Disarmament Affairs, Strategic Plans Division, Khalid Banuri, "Missile Defences in South Asia: The Next Challenge," *South Asian Survey*, 2004, p.197.
58. "Pak Strategic Missiles can Penetrate Proposed Indian BMD Shield: Gen Musharraf," http://www.pakistanidefence.com, July 26, 2005.

This assumption is believed to stem from Pakistan's first use nuclear posture that is premised on the need for rapid deployment. Therefore, the necessity to "undertake steps to produce, equip, deploy, man and exercise ballistic missiles with operational units."[59] These, however, may be kept at a low alert status, given that India has a no first use posture and Pakistan can assume a period of warning before a crisis builds up. Nevertheless, it does realise that time would be of essence in case of numerical inferiority.

Diversified Weapons Based on Highly Enriched Uranium and Plutonium Designs

Initially, Pakistan opted for the plutonium (Pu) route to nuclear weapons. With the natural uranium–heavy water moderated reactor, Karachi Nuclear Power Plant (KANUPP) becoming critical in 1971, Pakistan believed that by acquiring a reprocessing plant, it could access plutonium. This, however, proved to be easier thought than done. Not only did KANUPP prove to be inefficient and under International Atomic Energy Agency (IAEA) safeguards, but Pakistan's contract with France for the reprocessing plant was jettisoned by US non-proliferation concerns and stood cancelled in June 1978. Meanwhile, as a hedging strategy based on a conscious decision to accelerate its nuclear programme after 1974, Pakistan had already begun active pursuit of the uranium enrichment technology with a formal launch of a project the same year. In this, it met with far greater success since A.Q. Khan surfaced at just the right time with the stolen designs of centrifuges to procure enriched uranium,[60] and today Pakistan's nuclear weapons are primarily based on highly enriched uranium (HEU). The main enrichment facility exists at the KRL at Kahuta and is being supplemented with

59. Rodney W. Jones, "Pakistan's Nuclear Posture: Quest for Assured Nuclear Deterrence – A Conjecture", *Spotlight on Regional Affairs* (published by Institute of Regional Studies, Islamabad), vol 19, no. 1, January 2000.
60. For details on A.Q. Khan's operations, see Corera, n. 30.

a new enrichment facility near Wah (Gadwal uranium enrichment plant). Meanwhile, smaller, pilot scale enrichment facilities exist at Chaklala, Sihala and Golra.

Pakistani nuclear tests carried out in 1998 were of low yield, HEU fission weapons of varying designs, yield and sizes. Pakistani officials themselves have provided different yields of the five devices tested on the first day with some suggesting it to be of 25 and 12 kiloton (kt), besides three sub kiloton devices, and A. Q. Khan claiming that one of the devices was a boosted fission device of 30-35 kt, the other four being tactical weapons.[61] The sixth device, tested on May 30, 1998, is believed to have been a plutonium weapon, though there is no consensus on this.[62] In any case, according to some Western sources, Pakistan has at least two different basic nuclear weapon designs. The first developed by the Pakistan Atomic Energy Commission (PAEC) for air delivery has a range of about 10-20 kt and the other with a larger yield of about 15-25 kt is an HEU, implosion warhead design "of proven reliability" procured by Pakistan from China in the early 1980s.[63] It is known that China had also conducted a nuclear test for Pakistan in May 1990 at its Lop Nor test site. Pakistan has also devoted attention to the miniaturisation of nuclear warheads for making them light, compact and easily deliverable. Most of its missiles have a payload of 500 kg and the Pakistani nuclear warhead based on the Chinese design is expected to weigh about the same.

---

61. IISS Dossier, n.2, p. 33
62. The American Los Alamos Laboratory had claimed it to be a plutonium device on the basis of environmental sampling. But the report was contested by Lawrence Livermore Laboratory. It may also be recalled that after the North Korean nuclear test in October 2006, which was believed to be a plutonium weapon, President Musharraf had denied any Pakistani role since he claimed Pakistan did not have a plutonium weapon.
63. A.Q. Khan claimed as much in 1998. IISS Dossier, n.2, p. 32.

Tactical Nuclear Weapons (TNW)

Development and deployment of atomic artillery weapons has been denied by Gen Kidwai[64], but some others like Brig Saeed Ismat have written in favour of tactical nuclear weapons.

> We should have well defined and declared strategy of using our ultimate choice of nuclear weapons aimed at the destruction of those military forces, which have intruded in our territory. Our aim should be the destruction of the invading military forces only and not his civilian population. We should aim to strike with tactical nuclear weapons at the base of enemy offensive... Some standard artillery guns, rockets, and missiles can deliver these, so can helicopters and aircraft. Such low yield, high radiation nuclear weapons can quickly and decisively alter the entire course of battle. Though tactical in characteristics these can produce strategic effect.[65]

According to this viewpoint, given Pakistan's lack of geographical depth and inferior military resources, TNW are the only possible tools for tilting the balance in favour of Pakistan by threatening enemy intruders with a planned employment of these weapons. This strategy, in fact, has been propounded as being a *more moral one* and described as "Pakistan's Military Doctrine of Necessity". Also given that Pakistan projects its nuclear weapons as militarily usable, this "conventionalisation" of its nuclear weapon appears natural.

Credible Delivery Systems

Quite like India, the first nuclear delivery platforms available with Pakistan were aircraft. F-16 fighters[66] have been suitably

---

64. Cotta-Ramusino and Martinelli, n.4
65. Brig Saeed Ismat (Retd), "A Conceptual Nuclear Doctrine", *Defence Journal*, vol.3, no.8, March 2000.
66. Pakistan is currently estimated to have 47 F-16s in service. It awaits delivery of another 42 from the US and there is an option for 18 more to be obtained in the future, taking the total to 107.

modified to carry nuclear bombs, even as Mirage V and Chinese produced A-5 are also available.[67] The modifications, however, had the disadvantage of reducing speed and manoeuvrability. Therefore, even as weapon delivery flight training and bomb design modifications went on, Pakistan began to explore the option of procuring ballistic missiles (BMs). In any case, after 1985, under the Pressler Amendment further delivery of aircraft was impossible, and BMs were seen as the new potent symbols of credible deterrence.

A.Q. Khan, once again, proved successful in this field when he struck a deal with North Korea for the liquid fuelled No-dongs. Renamed Ghauri in Pakistan, versions of different ranges of this missile have been regularly tested for up to 1,500 km. Meanwhile, another programme run by the National Defence Complex with the Pakistan Space and Upper Atmosphere Research Commission (SUPARCO) and PAEC has concentrated since 1980s on solid fuel propelled short-range ballistic missiles, the Hatf series. Available in different ranges, these missiles are believed to be based on the Chinese M-11. However, it is the 290 km, single stage, solid fuelled and road mobile Ghaznavi or Hatf 3 which is declared to be nuclear capable and operational with the army from 2004 onwards. Pakistan also has a vigorous research and development (R & D) and procurement programme for medium range ballistic missiles (MRBMs) of a range of 650-1,000 km. Shaheen1 (600-800 km) from this category is nuclear capable and in service from 2003. Shaheen 2 (two stage, solid fuel) with a range of 2,000 km and carrying multiple warheads, and Ghauri 2 (liquid fuel) also

67. These will gradually be replaced by more modern J-10s and JF-17s that are being produced with Chinese help. Significantly, the US claims to have taken care not to provide Pakistan with any equipment that would facilitate nuclear delivery missions, such as the electrical mechanisms necessary for safe maintenance, transportation and delivery of nuclear weapons by F - 16s.

of about the same range, are still undergoing tests. The former is expected to provide the true deterrent to Pakistan while the latter would be an alternative. Over the years, Pakistani missiles have been equipped with better guidance systems, longer ranges and better accuracies. Babur, a 500 km, nuclear capable, cruise missile is also operational. Meanwhile, Ghauri 3 with a design range of 3500 km and possibly drawn from the North Korean Taepodong missile, is undergoing testing.[68]

Considering the active missile programme of the country, Pakistan's keenness to equip itself with a credible deterrent delivery capability based on mobility, dispersion and concealment is evident. It seeks through these measures to enhance the survival of its nuclear force against a disarming strike. Meanwhile the consistent pursuit of missiles of ever longer ranges may be explained for two reasons: one, to be able to hit deeper into India; and, second, to base its own missiles in more remote and less accessible mountainous areas in order to enhance their survivability and ensure second strike capability.

Accumulation of Fissile Material

Given the secrecy and opacity that normally shrouds nuclear developments, there are no confirmed figures available on Pakistan's fissile material stocks. According to Institute for Science and International Security (ISIS) estimates, at the end of 1999, Pakistan had 585-800 kg of HEU and 1.7-13 kg of separated Pu.[69] A more recent estimate, however, places Pakistan's HEU stockpile at 1,300 to 1,500 kg and 90 kg of Pu.[70] This is roughly expected to

68. For more detailed information on Pakistani missiles, see "Pakistan's Nuclear Forces, 2007," Nuclear Notebook prepared by Natural Resources Defence Council, *Bulletin of Atomic Scientists,* May-June 2007.
69. David Albright, "Securing Pakistan's Nuclear Weapons Complex," October 2001, http://www.isis-online/publications
70. Zia Mian, A.H. Nayyar, R. Rajaraman and M.V. Ramana, *Fissile Materials in*

translate into about 65 HEU weapons at 20 kg HEU per weapon and about 15 Pu weapons at 6-8 kg per weapon. However, there can be no direct derivation of the number of weapons from the fissile material stockpile since it is a more complex function of specific warhead designs, and their efficiency. Only some rough estimates are possible.

In any case, it could be said with some certainty that Pakistan is keen to accumulate as much fissile material as possible in order to increase its arsenal to roughly what India is believed to have. Its extreme discomfort at the thought of the FMCT freezing its numerical inferiority by proscribing future production of fissile material for weapons is evident in its approach to the treaty in the CD. Refusing to even call it a cut-off treaty since that has only a prospective connotation, Pakistan desires the treaty to deal with existing stocks as a way of reducing asymmetry. Key aspects of the Pakistani official position on the matter include[71]:

- Progressive reduction and eventual elimination of existing stocks of fissile materials.
- Schedule for transfer of stockpiles into civilian use with verification (transfers must first be made by the states with the largest stockpiles).
- Cap on future stocks to be accompanied by a reduction in asymmetries of existing stocks.

Asymmetry in fissile material stocks is a Pakistani concern and it would be natural to assume that it seeks to exploit its negotiating position to stall movement and buy time for accumulating as

South Asia: The Implications of the US-India Nuclear Deal, Draft Report for the International Panel on Fissile Materials, July 2006, pp. 11-15.

71. As cited in Gaurav Rajen and Michael Vannoni, "Fissile Materials Control in South Asia: Regional Analyses and Potential CBM," *Cooperative Monitoring Centre Occasional Paper,* February 2006, www.cmc.sandia.gov/links/cmc-papers/occassional-paper s

much fissile material as possible. It is revealing that besides HEU, Pakistan has also been looking at the Pu option afresh since its 50 MW reactor went critical at Khushab in 1998. Built with Chinese help and unsafeguarded, it is claimed to have the capacity to produce 10-15 kg Pu every year. An unsafeguarded pilot scale reprocessing facility is now operational at the New Labs at the Pakistan Institute of Nuclear Science and Technology (PINSTECH) complex near Rawalpindi. Meanwhile, a second reactor is being built at Khushab and has been variously estimated at having a capacity between 40-100 MW or even 1,000 MW.[72] If the latter estimate is true, the reactor could annually produce about 200 kg Pu that would be enough for 30-40 nuclear weapons a year according to an IISS estimate. However, before getting alarmed by such estimates that appear quite exaggerated, it must be pointed out that Pakistan does not yet have the reprocessing capacity for such amounts of spent fuel.

Uncertainties of Nuclear Deterrence

While the true value of deterrence lies in its remaining untested, or holding up in case of a crisis, the paradox is that for deterrence to be credible, one has to be prepared for it to breakdown. This possibility needs to be factored in by India with respect to Pakistan that has an offensive nuclear doctrine. In an approach best suited to its national interest, Pakistan has projected a carefully cultivated strategy of escalation spinning out of control if India were to launch a major conventional attack. Extremely intelligently, Pakistan has worked at reducing the risk of retaliation against its proxy war by heightening the risk of war if India were to press action. And, simultaneously, it has sought to reduce the risk of war by threatening that any war could turn into an all out, nuclear war.

72. IISS Dossier, n.2, p. 21.

In Western literature this has been described as the "risk maximizing approach" that relies on the enemy's fear that pressure exerted from its side could "provoke a viscerally violent response rather than a rationally restrained one."[73] By making nuclear threats it seeks to manipulate risks to its advantage even if following through on them would be nothing short of suicidal for itself. Rather bombastically, it has been stated by serving military men, "If Pakistan is being destroyed through conventional means, we will destroy them by using the nuclear option, as they say, if I am going down the ditch, I will also take my enemy with me."[74] Whether this would really happen or not, Pakistan banks on uncertainty bordering on desperation, or irrationality about its actions for deterring India.

Such a policy of brinkmanship that exploits uncertainty and rests on a perception of irrationality, however, carries the risk of deterrence breakdown due to misperception or miscalculation. There is no guarantee that the threshold of tolerance will never be breached or that the other side may not actually undertake an offensive and call Pakistan's bluff. In fact, if this were to happen, it would actually be in the best interests of Pakistan to avoid the use of nuclear weapon in order to retain the credibility of its nuclear deterrence for the future. It may be recalled that one of the reasons that the US did not use the nuclear weapon in the Korean War in the 1950s was the fear that if the bomb was used but did not achieve its purpose, then it would "undermine the very foundations of American strategic policy which rested so heavily on the weapon's omnipotence." Therefore, as the US discovered,

---

73. Thomas Schelling, *The Strategy of Conflict* (London: Oxford University Press, 1960).
74. Lt Gen Javed Ashraf Qazi made this statement to the Press in 2002 at the height of the India-Pak stand-off. Cited in Rahul Roy-Chaudhury, "Nuclear Doctrine, Declaratory Policy and Escalation Control," Stimson Centre South Asian Regional Security Project, April 2004.

the only way to maintain the credibility of the bomb was not to use it. "Deployment meant running the risk that its mysterious power might be revealed as a sham – that the Emperor might be discovered to have no clothes."[75]

However, there are a few scenarios where Pakistani nuclear weapons could come into play and these need serious consideration.

As a Result of Premeditated Attack

This could result from Pakistan mounting a conscious and planned, sufficiently large (mostly, but not exclusively counter-force) nuclear strike against India to end a conventional confrontation moving in India's favour. Faced by the prospect of military defeat, Pakistan would hope to wreak massive destruction in order to spread chaos and demoralise the Indian public and leadership enough to refrain from retaliation. It would also hope that in the remote possibility of an Indian nuclear response, it would be weak enough for Pakistan to handle since all major Indian delivery sites would have been destroyed. While this may be a dream scenario for Pakistan, fortunately for India, Pakistan does not yet have the capability to undertake such a strike. It cannot hope to obviate even a substantive part of the Indian arsenal, especially if India pays requisite attention to survivability. Pakistan would only end up inviting assured destruction in the bargain. Moreover, there can be no reason to assume that Pakistani decision-makers, civilian or military, could be so completely irrational. Every leader has a stake in the survival of his state because it is his source of power and hence, a deliberate suicidal mission by a rational leader appears improbable.

Unless, a radical Islamist leader (military or civilian) was in the seat of power. Indoctrinated with a suicidal mentality that

75. Degroot, n. 12, p. 188.

places little value upon his own survival or that of his nation, such a person could launch a pre-meditated nuclear strike on India in a bid to destroy the enemy, even at the cost of great sacrifice.

One could, however, debate the possibility of such an eventuality. Radical Islamist political parties have not been able to garner significant votes in elections. On the other hand, the rise to power of a radical military man through a military coup is possible. But, could he alone, in complete disjunction with the administration, be able to commandeer the country's nuclear arsenal? Nuclear infrastructure involves a large number of people who do impose a system of checks and balances. The SPD claims that Pakistan follows the three-rule principle for authorisation of nuclear use. Could all three people in the nuclear loop be equally motivated/deranged? Moreover, one must concede that under US pressure since 9/11, Pakistani nuclear command and control has matured from a personalised, ad hoc system into a more institutionalised one, that one hopes could not only survive regime change and domestic political upheaval, but also neutralise an irrational leader.[76]

As a Result of Accident or Miscalculation

This could occur in three ways.

- Firstly, as a result of a failure of the Pakistani command and control (C2) system. Given the small nuclear force and Pakistan's posture of first use, its C2 structure does require some delegation of authority. For reasons of survivability too, geographic dispersion of the small arsenal is a necessity, which brings its own challenges of effective and timely communication. Moreover, as launch authority flows

76. The Strategic Plans Division was created in 1999, the National Command Authority was set up in February 2000 and an Export Control Act passed in September 2004.

downward, the human factor becomes significant and any unauthorised launch of nuclear weapons, though remote, cannot be ruled out.

- Secondly, if Pakistan wrongly perceives its nuclear force as having been inadvertently destroyed by a conventional Indian counter-force strike, simply because it happened to be deployed in the region, it could result in nuclear use. Pressures of time in a crisis can lead to faulty decision-making without checking facts.
- Thirdly, as a result of an army misadventure. The last possibility is feasible given the past record of the Pakistani military in planning and authorising offensive ventures. In an attempt to distract attention from domestic unrest, or genuinely believing that it could pull off a strategic success against Indian conventional forces, deriving confidence from its recent large acquisition of modern conventional weapons, the army could be expected to indulge in such thinking. The Pakistan Army, in any case, has a high self-image as the defender of the nation. Unfortunately though, this does not detract from the conclusion that had been arrived at by Sun Tzu, "Military organisations that assume policy responsibility often make poor strategic choices." In fact, even the organisation theory identifies a set of parochial military interests and biases that lead senior officers to favour offensive doctrines, preventive wars, preemptive strikes, decisive counter-force options without thinking them through. Kargil was one such incident and the flawed strategic thinking that led to it has been criticised by many Pakistani analysts, including retired military men.[77]

77. For instance, Maj Gen Durrani observed that "Kargil was the result of flawed strategic thinking in Pakistan." Maleeha Lodhi attributes Kargil to "systemic flaws" in the Pakistani decision-making process which is " impulsive, chaotic, erratic and overly secretive." Maleeha Lodhi, "Anatomy of a Debacle," *Newsline*, July 1999, pp. 32-33.

And yet, there are several who blame the civilian government of the time for buckling under US pressure and denying the Pakistani military an opportunity to avenge past wrongs.

In Desperation or Despair

If the conventional military might of India were to endanger Pakistan's *survival as a nation*, Pakistan may use its nuclear weapon in desperation. This is perfectly understandable because nuclear weapons are measures of last resort. India could help avert such nuclear use by clearly articulating the military objectives of a conventional strike.It will have to be stated, as has been done in the past, that removing Pakistan from the face of the political map of the world is not an Indian political or military objective. Therefore, with Pakistan's survival not at stake, there should be little reason for a resort to nuclear weapons. Of course, this logic can work only if in Pakistani thinking the survival of its military regime is not equated with that of the nation-state.

Yet another reason for nuclear use could result from a moment of despair, if the Pakistani leadership finds itself in such a hopeless situation, domestically and internationally, that it finds *greater sense in self-annihilation than life after war.* Indications of this thinking were provided in an interview of a retired Pakistani general who claimed to speak for several others when he said that the situation in the country was so despondent as to merit a fresh start after a nuclear war!!

However, it needs to be pointed out to Pakistan, that despite the country being in dire straits—economically, politically, socially – the situation could be salvaged if it could reorient its priorities differently. Ever since independence, Pakistan's attention and energies have largely focussed on negating or eroding India's achievements – of democracy, secularism, economic development — by fomenting trouble and fanning insurgencies. A proxy war has been sustained through terrorism based on the radical ideology of *jehad,* physically supported through a network of training camps, and financially aided by narco-trafficking.

While India has been badly bled by such policies, it has not been possible for Pakistan to escape the repercussions of playing with fire either. Terrorist organisations have acquired a life and agenda of their own. The easy availability of weapons[78], and an illicit drug trade, as also distorted development priorities have led to a decline in social sector spending on public education so that *madrassas* remain the only option for many. With an unemployment rate of nearly 16 per cent, and *madrassa*-trained youth anyway unable to compete for modern jobs in the government or the limited private sector, joining *jehad* presents itself as a means of gainful employment for the majority. These trends, however, are reversible over time given requisite political will of the Pakistani leadership. The consequences of not acting quickly and decisively could be disastrous for the Pakistan state and already at the time of writing this chapter, the President Asif Ali Zardari had warned that the Taliban was trying "to take over Pakistan" even as the government signed an agreement with an ultra-Islamic extremist group to accept imposition of Islamic law in Swat Valley.[79] Developments such as these cause concerns for nuclear security too and are discussed in the next section.

---

78. It was estimated in 2004-05 that nearly two million small arms were in circulation in Pakistan while about three million rest in clandestine stockpiles available to a militant outfit ready to take up the cause of Kashmir.
79. "Pak Gives in to Taliban, Sharia is Law," *The Times of India,* February 17, 2009.

As a Result of Nuclear Use by Terrorists With/Without Government Support

One positive outcome of the revelations of the A.Q. Khan proliferation network was to compel Pakistan to enhance the security of its nuclear material, technologies and weapons in order to address Western fears of nuclear terrorism. Amongst the measures undertaken, the 5-year National Nuclear Safety and Security Action Plan initiated by the Pakistan Nuclear Regulatory Authority (PNRA) stands out for establishing a robust nuclear security regime that would minimise chances of theft of a nuclear weapon or fissile material. An official of the PNRA has claimed that "controls around various installations and radiation facilities in Pakistan are enough to deter and delay a terrorist attack...."[80]

However, three possibilities of nuclear terrorism with or without government support cannot be dismissed: one, that of a fundamentalist army general or a radical government in power itself clandestinely passing on a weapon to non-state actors for a strike against India. India should be able to deter such a possibility by stating a clear policy of treating this as first use of nuclear weapons and resorting to retaliation against the culpable state; second, Taliban or Al Qaeda terrorists acquiring nuclear material because of their links with some retired military and Inter-Services Intelligence (ISI) officials and nuclear scientists and using them to manufacture a radiological dispersal device (RDD);[81] and third, the possibility of a terrorist strike on, or sabotage of, Indian nuclear

80. Abdul Mannan, *Preventing Nuclear Terrorism in Pakistan: Sabotage of a Spent Fuel Cask or a Commercial Irradiation Source in Transport,* Stimson Centre, April 2007,. p.4.

81. This has been referred to as the most worrisome trend by a US South Asia analyst, Lisa Curtis, in her testimony delivered before the Subcommittee on the Middle East and South Asia and the Subcommittee on Terrorism, Non-Proliferation and Trade of the Committee on Foreign Affairs of the US House of Representatives on June 27, 2007. For text of the Testimony, see http:www.heritage.org.

facilities, including while spent nuclear fuel is being transported. Both these latter cases would cause massive contamination, panic, economic damage, and political instability besides degrading response capabilities and having a traumatic socio-psychological impact. India's response to such acts will primarily have to be enhanced intelligence to prevent, and better preparedness of its civil defence organisations to deal with, them. India has been preparing for the last few years to handle such scenarios with the institution of the National Disaster Management Authority. The terrorist acts would also, most likely, lead to more decisive action being taken across the LoC to hit out at terrorist camps and infrastructure. And, that's the time when nuclear deterrence will be tested.

Conclusion

The above analysis clearly indicates that Pakistan sees the nuclear weapon as insurance of its survival. What is extremely significant is that there is strong domestic public support for nuclear weapons. According to one analyst, "Workers and peasants, *maulvis,* white collars and intellectuals worship the Bomb" and even perceive it as a "great comforter, giving them spiritual strength to endure the hardships of life."[82] This support is largely a result of the conditioning of the domestic audience by the Pakistan Army that the country faces a serious security threat from India which can be met only if the army remains a major political decision-maker in the affairs of the country and retains a free hand with its nuclear policy. The latter, in fact, has been clearly projected as a military domain, with the army showing little tolerance for civilian interference. By all indications, the Pakistani nuclear programme will remain military driven until such time as an effective democracy can take root.

---

82. Artem Rudentsky, "Understanding Nuclear Pakistan: Global, Regional and Domestic Dimensions," *World Affairs,* vol.9, no. 3, Autumn 2005, p. 51.

Pakistan has an intelligent and rational nuclear policy best suited to its national security needs. Though bound by minimum nuclear deterrence, since a free run of resources is anyway not available, the first use nuclear doctrine makes a deliberate attempt to demonstrate irrationality and heighten the nuclear danger for two audiences: one, to instill fear in India that it has a low nuclear threshold; and second, to attract international intervention on Kashmir. Pakistani nuclear weapons play an important role in enhancing its defensive capability to meet the perceived threat from India and are designed to deter both conventional and nuclear aggression. In the process, Pakistani leaders of every hue have not shied away from making loud pronouncements on nuclear use, which according to some Indian analysts, have even followed a clear pattern – that of nuclear threats being louder during the beginning or end of the crisis and not during the crisis.[83]

However, doctrines alter with change in capabilities. How would conventional modernisation and growing nuclear and missile capability affect the country's nuclear doctrine? Would a better-matched conventional capability with India make Pakistan more, or less, restrained to undertake adventurous offensives? And, less or more prone to nuclear bluster? Would improved survivability of its nuclear arsenal make a no first use posture acceptable to Pakistan? In a normal nuclear dyad, this should have been so. But it appears less likely in the case of Pakistan since it is a revisionist nuclear power. Even though Agha Shahi had said that "the threshold of nuclear use will be inversely proportional to the level of balance or imbalance in conventional forces," since Pakistan seeks a change in *status quo*, the acquisition of modern conventional weaponry may not translate into nuclear restraint. In fact, despite technology and capability advancements, some nuclear facts are likely to remain unchanged:

83. For more on this, see Rajagopalan, n. 38, pp. 56-57.

(a) Nuclear weapons will remain central to Pakistan's national security.
(b) Till such time as Pakistan finds it prudent to abandon its proxy war against India, or give up its revisionist designs, it will retain first use doctrine.
(c) Despite large-scale modernisation of conventional forces, Pakistan will seek deterrence against an Indian conventional attack by projecting easy use of nuclear weapons.
(d) Given the continued centrality of nuclear weapons to its national security, Pakistani priorities in the next 10-15 years will be to improve survivability, penetrability and its command, control, communications, intelligence (C3I) capabilities. This will be achieved through creation of hard and deep buried storage and launch facilities, air defence around strategic sites, mobile missile units, concealment and deception, and possibly a strategic triad.
(e) Acquisition of missile defence to nullify India's advantage in this field and the simultaneous development of counter-measures would be undertaken, with significant Chinese help.

Since nuclear deterrence is essentially a mind game, Pakistan cannot be grudged any of the above actions. At the same time, India need not be out of depth in dealing with a nuclear Pakistan. India must understand the role of nuclear weapons in the Pakistani national security psyche and its dependence on them. This would free it from being self-deterred by Pakistan's nuclear bluster. At the same time, the dangers of nuclear escalation must be realised and treated with the seriousness they deserve. In addressing the nuclear threat from the Pakistani state, India must strengthen it own nuclear deterrence, paying special attention to the survivability issues and reliability and accuracy of delivery systems. And, for addressing the contemporary threat of nuclear use by non-state actors, India would do well to enhance its civil defence preparedness, and get the US to put pressure on Pakistan for greater safety and security of its nuclear assets, including further refinement of its command and control structures and processes. Taking the help of the United States in this matter would not amount to outsourcing India's security, as some allege, but to effectively making use of a leverage that India today enjoys and Pakistan does not.

3

Fire in the Dragon
China's Nuclear Doctrine and Strategy

Every nation possessing nuclear weapons is normally guarded in sharing information on related issues. This is even more true in the case of China that has over the last 2,500 years honed ambiguity, opacity and deception in matters of defence and military strategy into a fine art. Not surprisingly, therefore, Beijing officially offers little on its nuclear capability or strategy. Also, given that most domestic academic writing on the subject is in Chinese that few in the West and even fewer in India are fluent in, China's nuclear thinking and doctrine can at best only be surmised on the basis of rare[1], often rhetorical, and sometimes confounding official pronouncements or documents such as the White Papers on National Defence (WPND), translated articles in

1. Recent Chinese sources reveal that formal research and writing on doctrinal or operational nuclear issues began in China only in the mid-1980s. During the Cultural Revolution (1966-75), all academic military institutions were closed. It was only after 1978 and Deng Xiaoping's introduction of "reform and openness" that the intellectual hiatus ended. For more, see Evan Medeiros, "Evolving Nuclear Doctrine." in Paul J. Bolt and Albert Willner, eds., *China's Nuclear Future* (London: Lynne Rienner, 2006), pp. 44-49.

journals, scattered information on military capability build-up, and an understanding of the country's strategic culture, historical experiences and future ambitions. Of course, one abundant source on the subject comprises the American analyses by several government departments and individual China experts. But this literature is coloured by an American perspective that can only be of partial relevance to India, a country with which China has unresolved territorial disputes.

As a "rising China" rapidly extends its economic, energy and strategic footprint in the region and beyond, it inevitably casts a shadow on India. Of course, India is nowhere yet near China in economic growth, strategic military capabilities or the international security dispensation. But it is an aspirant to all of these, a reality that makes both nations potential rivals for the same resources and areas of influence. The present singular focus of China may be economic development to attain a well-off society by 2020, but there are apprehensions about how the country would eventually use its economic and military strength.

Officially, Chinese policy towards its neighbours claims to be one of "reassuring, benefiting and befriending" them.[2] With India too, bilateral relations appear more cordial than ever. However, occasional incidents in, and unfriendly references to, Arunachal Pradesh (only one of the disputed areas) serve as indicators of another kind, especially when seen in conjunction with China's ongoing military and strategic modernisation. Interestingly, its WPND 2006, outlines a 3-step modernisation strategy for the Chinese armed forces: "The first step is to lay a solid foundation by 2010, the second is to make major progress around 2020 and the third is to basically reach the strategic goal of building informationized armed forces and *being capable of winning*

2. Zhang Tuosheng, "China in New Phase of World Integration," *China Daily*, October 17, 2007.

informationized wars by the mid-21st century."[3] While the US may be the prime point of reference for most of these developments, the WPND does mention that "complex and sensitive historical and current issues in China's surrounding areas still affect its security environment."Statements such as these raise Indian threat perceptions, making it critical for New Delhi to monitor the changing nuclear mettle of a neighbour that is designated a potential adversary.

It is against this backdrop that this chapter performs five tasks: it analyses the role of nuclear weapons in China's security strategy; it studies the unique attributes of China's nuclear deterrence; it determines the likely impact of the US' RMA and deployment of BMD on China's nuclear posture; it identifies the trends in China's nuclear modernisation; and derives implications for India's nuclear strategy.

Purpose of China's Nuclear Deterrent

Given the historical circumstances in which China embarked upon its nuclear journey, the immediate purpose of the weapon was to impose deterrence against the use or threat of use of nuclear weapons. It may be recalled that the US had brandished nuclear threats against China in the 1950s in the context of the Korean War and the Taiwan Strait crises.[4] Therefore, *prevention of nuclear blackmail* and *safeguarding the country's independence and survival* were the primary motivations.[5]

3. White Paper on National Defense, 2006 as available on http://www.fas.org. Emphasis added.
4. In July 1950, at the very beginning of the Korean War, President Truman ordered ten nuclear configured B-29s to the Pacific. In 1952, President Eisenhower publicly hinted at nuclear use against China if it did not constructively participate in the Korean War armistice talks.
5. It is a different matter that with growth of its nuclear capability, China has not hesitated to indulge in a bit of nuclear blackmail itself. One need only recall the threats made in early 2000 to proliferate nuclear and missile

Efforts in this direction started in 1955 with the establishment of the Ministry of Nuclear Industry and with help from the USSR. During 1955-58, China was the recipient of Soviet nuclear largesse by way of assistance in development of its nuclear industry and training of scientists.[6] However, in June 1959, the Soviets reneged on their commitment to provide Beijing with a nuclear weapon prototype[7], compelling it to launch its indigenous efforts. Demonstration of domestic scientific prowess became an issue of national and international prestige and China's ability to overcome technical hurdles independently and detonate a nuclear bomb in 1964 significantly raised its stature, not the least in its own eyes. This confidence received a further boost when within 32 months of its first explosion, China successfully tested the H-bomb on June17, 1967.[8]

According high scientific and national importance to its nuclear programme, the Chinese leadership kept it insulated from the adverse impact of the Cultural Revolution and used it to build a solid foundation for scientific and technological research and development. It became, and remains, the symbol and substance of Chinese strategic autonomy. In Deng Xiaoping words, "If we had not possessed nuclear and hydrogen bombs, and launched satellites since the 1960s, China would not be regarded as an

technologies in order to counter the US decision to proceed with missile defence. In any case, China has used proliferation as a means to balance power, as its nuclear and missile transfers to Pakistan illustrate.

6. It is documented that "six Sino-Soviet nuclear accords were signed between 1955-58, ranging from joint uranium prospecting to the transfer of Soviet nuclear weapons technology," which even envisaged the supply of prototype atomic bombs. For more on this, see Arvind Virmani, "Trilateral Nuclear Proliferation: Pakistan's Euro-Chinese Bomb," *IDSA Monograph Series*, no.1, December 2006, p. 28.
7. Interestingly, China's first nuclear device was code-named 59-6 for the year and the month (June 1959) of Soviet loss of support.
8. The USA had taken 86 months, the USSR 75, the UK 66 and France 105 to perform the same feat.

influential world power, nor would it have the international position it occupies today." Certainly, Beijing exploits its nuclear capability as a currency of power and prestige to secure its "rightful place" in the global system.

So, what Mao had dismissed as a "paper tiger" in 1946 when China did not have the weapon, did, after 1964, become an important counter-coercion weapon in the country's security strategy, and a symbol of great power status. Today, however, given that Beijing no longer perceives a threat to its 'survival', the *raison d'etre* of its nuclear weapons has expanded to safeguard additional national priorities such as economic development and eventual national reunification.

In the context of economic development, China's nuclear weapons perform the symbolic role of power projection to help ensure a stable security environment necessary for its continued economic growth, and concomitantly for its emergence as a great power. But these weapons come into their own as a powerful deterrent with regard to China's claims over territories that it believes rightfully belong to it. This has been most plainly articulated in the context of Taiwan. WPND-2004 bluntly states, "It is the sacred responsibility of the Chinese armed forces to stop the 'Taiwan independence' forces from splitting the country."[9] In fact, China perceives a real threat from US support for Taiwan's forces of independence and is wary of American arms influx into the region, especially the deployment of theatre missile defence (TMD). In order to obviate chances of American involvement in a crisis over Taiwan, nuclear weapons constitute an important component of China's deterrence strategy. Several analysts have not hesitated from suggesting, "If China's conventional forces are devastated, and if Taiwan takes the opportunity to declare *de jure* independence,

---

9. *China's White Paper on National Defence, 2004,* as available at http://www.fas.org

it is inconceivable that China would allow its nuclear weapons to be destroyed by a precision attack with conventional munitions, rather than use them as a true means of deterrence."[10] Maj Gen Zhu Chenghu, dean of the People's Liberation Army's (PLA's) National Defence University, also expressed this in July 2005, when he opined that Beijing could be tempted to abandon its no first use policy if faced with the prospect of intervention of the superior US conventional forces in a conflict over Taiwan. Of course, this was eventually, though not immediately, officially dismissed as the personal opinion of the General, there is no doubt about Beijing's sensitivity on Taiwan and its use of all military means at its disposal to deter outside intervention.

Beyond Taiwan, and in a more regional Asia-Pacific context, China sees its nuclear capability as a means of "guaranteeing that its interests are not undermined by a hegemonic United States seeking to impose 'a new world order' on other nations."[11] On the one hand, with the end of the Cold War, China perceives a greater threat from the fact that the American and Russian nuclear forces are no longer solely focussed on one another, "thus, freeing up more weapons for use against China."[12] This is especially true of the US that, by its own admission, allocates an extensive role to China in its nuclear planning. In fact, in February 2006, the US Office of the Secretary of Defence (OSD) identified China as the top-most large-scale military threat facing the US.[13] Meanwhile, Chinese strategists also consider their nuclear capabilities as a hedge against any reversals in Russian politics, or the possibility of US-Russian

10. Shen Dingli, "Nuclear Deterrence in the 21st Century," *China Security*, issue 1, Autumn 2005. p. 13.
11. Lisbeth Gronlund, David Wright and Young Liu, "China and a Fissile Material Production Cut-off," *Survival*, vol. 37, no. 4, Winter 1995-96, pp. 152.
12. Ibid., p. 152.
13. Hans M. Kristensen, Robert S. Norris and Mathew G. Mckinzie, *Chinese Nuclear Forces and US Nuclear War Planning* (Washington DC: FAS and NRDC, November 2006), pp. 18-19.

collaboration on missile defence. With the overt nuclearisation of North Korea, China also has to contend with increasing US-Japan and US-South Korea strategic and BMD alignment. Simultaneously, the evolving strategic relations between India and the USA (that ample American literature describes as aimed at "containing" China) do not help ease Chinese threat perceptions, and rather raise preoccupations on the possibility of India becoming a part of the future balance of power system in Asia. Consequently, India's military build-up and its potential growth that would enable it to effectively undertake regional and global engagements cause discomfiture in Beijing.[14] Placed in such a security conundrum, it is not surprising that China, quite like the other NWS, perceives its nuclear weapons as *insurance in a multi-dimensional nuclear environment* and against an uncertain future.

Nature of China's Nuclear Deterrence

It is remarkable that China defined and evolved its unique brand of nuclear deterrence at a time when the only predominant trend was the one being followed by the two superpowers that emphasised accumulation of large numbers of nuclear warheads placed on hair trigger alert to cause mutual assured destruction. And even more significantly, it did so at a time when the nuclear arsenals of both the superpowers were not only aimed at one another, but also individually at China. Despite this, Beijing rejected a large nuclear arsenal or low level of launch readiness. It settled instead for a *defensive second strike role* for its small nuclear forces, their size being just enough to cause "unacceptable damage" to the adversary. These attributes of nuclear deterrence were an outgrowth of Mao's broader military thought and based on no

14. Chinese assessments of India's future development and role are found in Michael Pillsbury, *China Debates the Future Security Environment* (Washington DC: National Defence University Press, January 2000).

formal, systematic research on nuclear operations and strategy, as had evolved in the US. Several decades hence, little more has emerged by way of amplification of this general and ambiguous set of beliefs.

Strategic ambiguity and deception, core components of Chinese strategic culture from the times of Sun Tzu[15], are important aspects of China's nuclear deterrence too. China has never proffered clear indications about the state of its nuclear arsenal, warheads or delivery capabilities, not even by way of endorsing or denying others' suggestions. However, there is a difference between the Chinese nuclear posture of the 1980s and now. While earlier, nuclear deterrence was solely based on ambiguity and secrecy in numbers and capability (indicating greater reliance on existential deterrence), it is now more credibly derived from mobility, invulnerability and penetrability of nuclear forces.[16] This change is the result of both China's perception that it has been accommodated in the outside world, as well as an increase in the number, variety and quality of its nuclear forces. Both imbue it with greater confidence and, hence, allow more transparency at one level, even as ambiguity on nuclear command and control, force posture, targeting philosophy, etc. remains. Based on available literature, however, the following salient features of China's nuclear doctrine stand out.

---

15. Amongst his oft-quoted writings on the subject is: "All warfare is based on deception. Hence, when able to attack, we must seem unable; when using our forces, we must seem inactive; when we are near, we must make the enemy believe we are far away; when far away, we must make him believe that we are near." Sun Tzu, *The Art of War,* trans Thomas Cleary (Boston, MA: Shambhala, 1999).
16. For more on this, see Medeiros, n.1, p. 54, and Jonathan Pollack, "The Future of China's Nuclear Weapons Policy," in John Hopkins and Weixing Hu, eds., *Strategic Views from the Second Tier: The Nuclear Weapons Policies of France, Britain and China* (La Jolla, CA: University of California, 1994), p. 163.

Political Use of Nuclear Weapons

Most Chinese writings, quite like in India too, perceive nuclear weapons as a political tool of deterrence. This is not surprising given that China has never had the capability to engage in nuclear war-fighting with either the US or USSR. It, therefore, proclaimed a defensive role for its nuclear weapon, projecting it as an effective tool that would be used to impose a high cost on the aggressor so that no gains appear worthwhile. Deterrence, therefore, is the primary purpose of the weapon. Sun Tzu's formulation on deterrence too had advocated "forcing the other party to resign to our will without fighting a battle."

Implicit in this form of deterrence is the communication of the Chinese resolve or willingness to respond to nuclear threats or use. For instance, the Science of Military Campaigns, instructs the Second Artillery to be "ready to fight while deterring" because "comprehensive and firm combat readiness is itself an important means and firm backing to show a strong resolve and will."[17] The fact that China's nuclear weapons were and are primarily meant to deter the USA makes it easier to exploit this political role of the nuclear weapon since the developed country, in Chinese perception, is not credited with a high damage tolerance threshold. The Chinese are almost smug (based on their reading of the Cuban missile crisis) that the US would never accept the loss of even one American city, no matter what their objective, whether in Taiwan or elsewhere.

With the deployment of American national and theatre missile defences, however, China has had to recalculate its ability to cause 'unacceptable damage' to the USA. It is consequently engaged in a process of strategic modernisation. Yet, it might still be said that *imposition of nuclear deterrence and rejection of nuclear war-fighting* would remain the primary objectives of the Chinese

17. Medeiros, n.1, p. 66.

nuclear doctrine, given that despite build-up of new capabilities, and a likely reduction in the American/Russian nuclear arsenals, China would still not be able to fight and win a nuclear war with either of them. Therefore, the focus of the Chinese nuclear strategy would remain "containment of nuclear war, not on winning a nuclear war."[18] China, like India, understands that it would be best to exploit the *political* benefit of the nuclear weapon, albeit by suitably enhancing the military aspect of its nuclear capability. China would prefer to keep hostilities conventional, though not hesitating to brandish a palpable threat of use of superior nuclear capabilities.

Minimum Nuclear Deterrence

The use of the nuclear weapon as a political tool of deterrence obviates the need for nuclear superiority, or even parity. Consequently, China has aimed for a capability that would be sufficient to deter threats of nuclear use against itself. In fact, Beijing has not bothered about the nomenclature of such a form of deterrence. Mao, in keeping with the Chinese strategic culture of minimalism, defence and ambiguity, directed his country to steer clear of an excessive build-up of nuclear warheads in favour of an arsenal that would be just enough to inflict unacceptable damage. It was Mao's dictum to "build a few, keep the number small, make the quality high" and he is also cited to have stated "Six are enough."[19] In Western literature, this was recognised as minimum nuclear deterrence.

Mao's rationale for such a doctrine was as much a result of circumstantial resource constraints of funds and fissile material, as the astute understanding that Washington would never risk

---

18. Brad Roberts, "China and Ballistic Missile Defence: 1955 to 2002 and Beyond," *Proliferation Papers*, IFRI Security Studies Department, Winter 2004, p.41.
19. Litai Xue, "Evolution of China's Nuclear Strategy" in Hopkins and Hu, eds., n.16, p. 171.

a nuclear exchange with China. Rather, Beijing found greater wisdom in setting an "example of moderation and prudence on the moral high ground, and seemingly proved its theory that small defensively oriented arsenals at once provided deterrence, reassurance and stability."[20] Mutual vulnerability was believed to be good enough and China never strove for MAD. The emphasis was on building high yield nuclear weapons for counter-value targeting, so that destruction was assured even without accurate delivery systems.

However, it was only to be expected that with economic and techno-scientific development, and change in threat perceptions and core interests, the nature of China's nuclear deterrent too would alter. Consequently, the Chinese development or acquisition of newer strategic capabilities is driven as much by the new threats (such as BMD), as by the logical progression of its economic prowess and techno-scientific expertise. Some Western Sinologists describe this as graduation from minimum to *limited nuclear deterrence*. The essential difference lies in the fact that while *minimum* only required China to maintain a credible second strike capability based on survivable nuclear forces, *limited* deterrence entails the possession of more sophisticated nuclear forces capable of controlling escalation during a conflict and bringing about intra-war deterrence.[21] This envisages flexible nuclear responses, including counter-force targeting, instead of basing deterrence on only counter-value punitive strikes. The emphasis, therefore, shifts to acquisition

20. Bruce G. Blair, "General Zhu and Chinese Nuclear Preemption," *China Security*, issue 1, Autumn 2006, p. 16.
21. Alastair Iain Johnston, "Prospects for Chinese Nuclear Force Modernization: Limited Deterrence Versus Multilateral Arms Control," *China Quarterly*, June 1996, pp. 552-558; Alistair Iain Johnston, "China's New 'Old Thinking': The Concept of Limited Deterrence," *International Security*, vol. 20, no. 3, Winter 1995/96.

of "limited war-fighting capabilities, improved command and control and early warning systems, smaller, survivable, mobile, more accurate and diverse cruise and ballistic missile nuclear delivery systems, possible abandonment of the NFU policy, and missile defence.[22] Some other analysts, however, such as Evan S. Medeiros, have described these moves as being driven by "*sufficiency and effectiveness.*" This reflects the need to build a force sized and scaled to resist enemy efforts to coerce China by threatening retaliation, and in case that fails, to survive an enemy's initial strike, to execute counter-attack and re-attack operations to inflict unacceptable damage, and to penetrate whatever defences the enemy may employ.[23]

Irrespective of the nomenclature for its deterrence, China is certainly graduating to a higher level of strategic capability given that it now has the economic wherewithal to transcend 'minimum', and also the political context in the deployment of national missile defence (NMD) in the USA, and theatre missile defence (TMD) in the theatres most relevant to Chinese security (Japan and Taiwan). The focus is to possess a *credible deterrence* with acquisition of relevant capabilities – the fact that these could allow a limited scope for nuclear war-fighting, especially with weaker nuclear powers to impose intra-war escalation control, will be used to further the deterrent value of the weapon.

No First Use (NFU)

Amongst the five nuclear weapon states recognised under the nuclear Non-Proliferation Treaty (NPT), China is presently the

22. Bates Gill, James Mulvenon, Mark Stokes, "The Chinese Second Artillery Corps: Transition to Credible Deterrence," in Mulvenon, Andrew N.D. Yang, eds., *The People's Liberation Army as an Organization: Reference Volume V* (2001), p. 548-549.
23. Medeiros, n.1, p. 41.

only country that has a declared NFU doctrine.[24] Amongst other concepts, China owes this too to Mao who crafted it as the best response to larger nuclear threats facing China. While, on the one hand, he used the weapon to project a definite threat of retaliation that would cause unacceptable damage, on the other, through NFU, he yet again claimed the moral high ground by projecting a measure of reassurance by taking the pressure off the adversary.[25]

Several Western analysts have described China's NFU as making a virtue out of necessity since it never had an extensive and advanced early warning system or hardened C3I infrastructure that could have accommodated launch-on-warning/launch-under-attack (LOW/LUA) postures to enable first use anyway. This viewpoint would be tested in the coming decades as China develops a more modern and sophisticated nuclear arsenal that would be capable of mounting preemptive nuclear offensives, while retaining a credible second-strike capability. Could it then abandon NFU? Indications of this thinking are evident in Chinese writings and 'personal' statements of military men, primarily in the context of new threats to the survivability of Chinese nuclear forces as a result of advances in US BMD and non-nuclear precision strike capabilities.

Though Party officials assert that the country's nuclear doctrine remains unchanged, one cannot discount the possibility that in case of a crisis over Taiwan, in which US involvement overwhelms China with its superior precision-guided weaponry delivered with the help of real-time intelligence, Beijing could be tempted to resort to first nuclear use in the theatre of war (not at the US

24. Russia abandoned its NFU position in 1992 once it recognised the weakness of its conventional forces.
25. Immediately after its first test, China launched a "Nuclear Peace Offensive" that included the NFU, a proposal for a ban on first use, and recommendations for a multilateral conference to discuss abolition and destruction of nuclear weapons.

mainland) to send a message to the US to stop its interference in the region. This temptation might grow with greater confidence in the mobility and survivability of its nuclear infrastructure that may push China to believe that it could impose intra-war deterrence. Beijing has intelligently invested in increasing the survivability of its nuclear arsenal and does not envisage a scenario where any adversary, not even the US, could conduct a disarming first strike. In any case, there are references in Chinese military writings to the use of demonstration strikes, with conventionally armed nuclear missiles, to signal China's resolve.[26] A counter-force nuclear strike with tactical nuclear weapons might well be justified as "counter attack in self-defence," a concept that has been upheld in WPND, 2006 and 2009, and was used earlier for other wars as well, including the one with India in 1962. A RAND Study of 2005 concluded that where use of force involves core interests such as sovereignty or territorial claims, Beijing "could claim military preemption as a strategically defensive act..."[27] Officially, of course, nothing is forthcoming, but some writings suggest that Chinese thinkers are grappling with how to respond to conventional attacks on strategic systems, or in case of an intelligence warning of imminent strategic attack. Could preemptive attacks in such cases be justified as "self-defensive counter-attacks", in keeping with China's "active defence" strategy?

Implications of any abandonment of NFU for India will be considered in the last section of the chapter. Meanwhile, vis-a-vis the USA, pursuance of a declaratory NFU posture is, and will remain, the best course of action for Beijing. It would make little strategic sense to alter it, at least at the rhetorical level, since NFU does make China appear less belligerent, more internationally acceptable, and even benign! It is also debatable whether first use

26. Medeiros, n.1, p. 61.
27. Kristensen et al, n. 13, p. 6.

would actually serve any purpose, and not prove to be suicidal instead. It would be better for China to enhance its nuclear deterrence through modernisation of its conventional forces, in order to keep the nuclear threshold as high as possible.

Meanwhile, if the international community (particularly the USA) is serious about dealing with possible nuclear first use, it may be advisable to take up China's often-made offer to evolve a universal NFU treaty, thereby increasing the spectre of international opprobrium on anyone who resorts to the first use of nuclear weapons. For a country like China that lays great score on its international image, this might constrain any temptation for nuclear use.

Tactical Nuclear Weapons (TNW)

It is widely believed that China has 150 tactical nuclear weapons that were mostly developed in the 1980s. Prior to that, Beijing stuck to Mao's dictum to build a few strategic thermonuclear weapons that could cause unacceptable damage even if delivered by inaccurate missiles over urban areas or soft military targets. But by the early 1980s, PLA leaders had concluded that enemy TNW could be deterred only by acquiring similar weapons, particularly to avoid the dilemma of either using "strategic weapons prematurely" or remaining paralysed. So, through the 1980s, China developed the neutron bomb and tested other low yield weapons. Simultaneously with the development of the 600 km range DF 15/M 9 missile, they had an effective means to employ nuclear weapons in theatre operations by the end of the decade. According to the *Military Balance*, Beijing had several hundred of other nuclear devices such as nuclear artillery shells, nuclear-tipped rockets, and nuclear demolition mines by the late 1990s, with plans to build hundreds more and to be mounted on solid-fuelled SRBMs and MRBMs, capable of carrying nuclear or non-nuclear warheads. *Jane's* publications too indicate that the People's Republic China (PRC) has nuclear mines for use at sea, nuclear torpedoes, nuclear depth charges, and nuclear anti-ship missiles.

Impact of Changing Conditions on Nuclear Posture

Among the five "musts" outlined by Jiang Zemin in 2000, the fifth one[28] was that "China must pay attention to the global situation of strategic balance and stability and, when there are changes in the situation, adjust its strategic nuclear weapon development strategy in a timely manner." It is in keeping with this commandment that China is today adapting its nuclear capability and strategy to the changing realities. Fortunately for the country, it is having to do so at a time when it can well afford the development and/or acquisition of modern systems and capabilities. Therefore, while the international context may be the prime driver for strategic modernisation, domestic economic and technological growth are the facilitators.

In the 1980s, economic development took precedence over military spending in China. But once the economy prospered, with a gross domestic product (GDP) growth rate of 9 per cent in 1978-2003, the country could support an increase in the military budget, officially stated by the US to be as much as 40 per cent in real terms over the last few years.[29] China's official data on defence spending (that does not include many

28. The other four musts included: must own strategic nuclear weapons of a definite quality and quantity; must guarantee safety of strategic nuclear bases against loss of combat effectiveness from attacks and destruction by hostile countries; must ensure high degree of war preparedness; must be able to launch nuclear counter-attack and nuclear re-attack against the aggressor.
29. According to Chinese sources, China's defence budget for 2006 is about $ 30 billion, though the US Defence Intelligence Agency places it at anything between $ 75-105 billion. Kristensen et al, n.13, p. 7.

crucial areas, including expenses for strategic forces, R&D, etc) indicates an average growth rate of about 14 per cent annually in real terms over the past 15 years. Whatever the real figures may be, the essential point is that China definitely is able to spend much more on defence than earlier. In fact, it may even be said that in the past, China's strategic force structure was most likely not necessarily determined only by doctrinal precepts, but also by economic and technological factors. Therefore, as these limitations lift, force modernisation is a natural corollary, particularly as the country grapples with the following specific threats.

Two major developments in contemporary US military posture cause particular concern to China. The first of these relates to the regular and perennially ongoing RMA. Beijing has particularly taken note of the new American triad[30], two of whose legs – deployment of nuclear and non-nuclear strike capabilities, and active and passive defences – directly impinge upon its deterrence. Both are enabled by improved surveillance and reconnaissance systems that widen the gap of military capabilities between the US and others. The Gulf War and Iraq War have amply demonstrated US ability to effectively use high technology, and integrated conventional forces to accurately destroy fixed targets. Beijing is extremely conscious of the already existing US conventional dominance through the use of digitally guided, space controlled precision conventional strikes that could not only cripple its conventional capability but also target immobile Chinese nuclear and missile storage or launch sites. As stated by one Chinese analyst, "Nuclear retaliation capability is threatened not only by nuclear forces but also by conventional forces... [which] may impair the strategic stability and induce a round of the nuclear

30. For more on this, see Chapter 1 of this book, section on "Blurring lines between conventional and nuclear weapons."

arms race."[31] While well conversant with the Soviet experience, and, hence, not wanting to overspend on defence or indulge in an arms race that could send alarming signals to the region, now an important trade partner and contributor to Chinese economic development, Beijing, nevertheless, is keen to design and deploy systems that would be difficult to identify, track and strike in a first or a follow on attack.

China's confidence in its counter-strike capability to cause unacceptable damage to US cities is further affected by the deployment of an effective BMD system. Therefore, US abandonment of the Anti-Ballistic Missile (ABM) Treaty in June 2002 and Russian overtures towards shoring up its own defences have caused much anxiety. In fact, China has objected to missile defence ever since the 1980s, when the Strategic Defence Initiative (SDI) was first proposed. Through the late 1990s and early years of this decade, as the debate revived in the US, China sided with Russia in its criticism of the American moves. However, it had realised its own inability (as well as Russian impotence) to halt US plans by the time America withdrew from the ABM Treaty. Reconciled to its inevitability, China had also by then begun to craft its own responses along two main trajectories – politico-diplomatic and military. Both have since been followed with finesse and tact so as not to antagonise the US with too strident an opposition[32], while also being prepared since it has not accepted US reassurances that the BMD was not aimed at it.

Beijing has used every international forum and opportunity to highlight the adverse impact of the US move on international strategic stability. Highlighting the "offensive" uses of the BMD system such as "accurately intercepting high-speed missiles

31. Qui Yong, "Preliminary Study on the Threat of Precision Strike Conventional Weapons to Nuclear Weapons", as cited in Roberts, n. 18, p. 40.
32. 16th Congress of the Chinese Communist Party in 2002 had committed to the maintenance of positive relations with the US as top priority.

attacking aircraft in flight and satellites in orbit," Foreign Ministry spokesmen have regularly stated that "missile defense systems negatively impact strategic balance and stability and are detrimental to mutual trust among major powers and regional security [and] will also probably trigger new proliferation issues."[33] Beijing has also been especially critical of US plans to deploy elements of its defence system in Europe, fearful that its success would fuel missile defence cooperation with Japan in Asia, a development now well on its way. However, realising its inability to influence the American BMD trajectory, and rather the likelihood of more nations going down that path, China is developing a more military response, as evident in its efforts at strategic modernisation.

Trends in Chinese Strategic Modernisation

Towards the end of the 1990s, the US Department of Defence undertook an assessment of the ability of some countries to produce advanced equipment such as command, control, communications, computers, intelligence, information(C4I2) systems; high-performance computing; intelligent systems; signal processing; transmission systems; opto-electronics, etc. It rated countries' capabilities on a scale of zero to four. While the US was at four, China received one, indicating a minimum level of proficiency.[34] Indeed, China is well aware of its limitations in these spheres, as well as of their criticality in wars of the future. The focus, therefore, is on "informationization" in its WPND 2004, 2006 and 2009. The last two, in fact, specifically called upon the Second Artillery Corps to "build a streamlined and effective

33. Joseph E. Lin and Christopher Griffin, "Cold War Redux: China Responds to the Russo-American BMD Dispute," *Jamestown Foundation's China Brief*, 11 July 2007.

34. As cited by Thomas M. Kane, "Dragon or Dinosaur: Nuclear Weapons in a Modernizing China," *Parameters*, Winter 2003-04, p. 101. Japan received a rating of four and India was at one.

strategic force by raising the informationization of its weaponry and equipment systems, build an agile and efficient operational command and control and increase capabilities of land-based strategic nuclear counterstrikes and precision strikes with conventional missiles." Evidently, the White Papers indicate an intelligent prioritisation of capability development to enhance Chinese security and exploit American vulnerabilities. Some of the specific efforts being undertaken by China to enhance the credibility of its nuclear deterrence are highlighted in the following paragraphs.

Increase in Size of Nuclear Arsenal

The most obvious response to BMD would be to increase the number of warheads and strategic missiles to saturate and defeat the missile defence. However, China remains absolutely opaque on the numbers except to state that it has the smallest arsenal among all the five NWS. Estimates of the numbers of its nuclear warheads vary from 200 to 450. For instance, the *Bulletin of Atomic Scientists* derives numbers on the basis of delivery vehicles to calculate that China deploys approximately 176 warheads (about 121 on land-based missiles, none yet on SLBMs and 55 on nuclear capable aircraft). Along with some additional warheads in storage, the total is conservatively placed at 240.[35] Some others, though, report about 300 strategic and 150 tactical nuclear weapons. Meanwhile, a leaked Defence Intelligence Agency (DIA) estimate from 1999 stated that China's total nuclear inventory would increase to 358-464 warheads by 2020.[36] Another source reveals plans to deploy 500 nuclear warheads by 2015, with 300 on MIRVed ICBMs and the rest on

35. Robert S. Norris and Hans Kristensen, "Chinese Nuclear Forces, 2008," *Bulletin of Atomic Scientists*, vol.64, no.3, July-August 2008, pp. 42-44.
36. Kristensen et al, n.13, pp. 35-40.

SLBMs.[37] US officialdom also predicts an increase of ICBMs from the present about 20 to 75-100 by 2015, some of which would have multiple warheads.[38]

While these figures can only be guesstimates in the absence of any official data, that there will be a build-up of the arsenal should be expected since China cannot afford to let improving US defensive and offensive capabilities degrade its deterrent. It would want to restore the status quo ante with a quantitative and qualitative increase in its own arsenal. In any case, this appears imminent given the increased financial resources, and a reinforced commitment to the Second Artillery that is reported to have "gained higher prominence in Chinese defence planning and decision-making over the last decade. Its doctrine has been thoroughly updated, reflecting enhanced reliance on missile warfare by the PLA more generally."[39] Although the Second Artillery makes up only 4 per cent of the PLA, it receives 15 per cent of the defence budget and 20 per cent of the total procurement budget.

Besides money and political commitment, any substantive increase in the number of warheads would also require adequate fissile material, which according to some sources, should not really be a problem. In fact, according to the National Resource Defence Council estimates, China should have enough fissile material stockpiles to double or even treble its nuclear warheads and even the RAND Corporation finds it feasible for the warheads to increase to anything between 600-1,500. Several Western scholars have

---

37. Centre for Non-proliferation Studies, Monterey, "China's Nuclear Weapon Development, Modernization and Testing," *China Profiles*, http://www.nti.org/db/china.
38. Ibid., p.2 .
39. Brad Roberts' book review of Jeffrey G. Lewis, "The Minimum Means of Reprisal: China's Search for Security in the Nuclear Age," in *Arms Control Today*, May 2007.

circulated the information that China had ceased production of HEU in 1987 and plutonium in 1991[40], even though it has never accepted or denied the veracity of such an assumption. However, it would be naïve to believe that the country would have mothballed its production facilities in any such way that resumption of production, assuming it was considered necessary, could not be undertaken. In any case, it could be presumed with certainty that China would join an FMCT only after it has secured itself on fissile material.

Improvements in Mobility and Survivability of Delivery Systems

Concealment to enhance survivability was always central to Chinese nuclear deterrence. In the absence of mobile missiles, the Second Artillery built several launch sites near each cave housing missiles and the commander was trained in guerrilla tactics to move the missiles around. Subsequently, with mobile missiles too, combat units practise random rotation of missiles among launch sites, moving them along different land routes and maximising the use of camouflage and night operations. More recently, the Second Artillery is believed to have focussed on training and readiness of operational units, rapid reaction, electronic warfare (EW), and precision attacks. Exercises to test reliable communications, combat coordination, damage control, equipment repair and rapid launch have been conducted.[41]

In order to enhance the survivability, accuracy and mobility of its strategic forces, China has been engaged in transcending to more solid fuelled, rail and road mobile missiles. China has begun deploying a new generation of solid fuelled mobile missiles, the DF-21A which is credited to have global positioning system (GPS) or radar-based terminal guidance for greater accuracy. Meanwhile,

40. Blair, n. 20, p. 19.
41. Office of the Secretary of Defence, *The Military Power of the PRC, 2005* (Washington DC: OSD, 2006).

still under development are the DF-31 and DF-31A, ICBMs that will form the core of its future strategic land-based missile force. The DF-31, flight tested since 1999, is likely to be a single warhead missile, but equipped with decoy warheads and other countermeasures to penetrate US BMD.

However, China realises the vulnerability of land-based missiles despite the best survivability measures. As was stated by Admiral Liu Huaqing, "In the face of a large scale nuclear attack, only less than 10 per cent of the coastal launching silos will survive, whereas submarines armed with ballistic missiles can use the surface of the sea to protect and cover themselves, preserve the nuclear offensive force and play a deterrent and containment role."[42] Hence, the focus on sea-based deterrence for greater survivability. The sea leg of the Chinese nuclear triad would credibly rest on Julang 2 (JL-2), a second generation SLBM that would be deployed on an indigenous Type 094 submarine that is expected to become operational by the end of this decade. One Xia class SSBN with 12 JL-1 missiles of 2,150 km range has been operational since 1988. But given its problems of high noise levels, radiation leaks and the ability to carry only short range ballistic missiles (SRBMs) with single warhead, it has never been considered a viable second strike deterrent.[43] New inductions are, therefore, considered critical for deterrence based on a counterstrike doctrine.

Meanwhile, 100 H-6 bombers capable of carrying three bombs over 3,100 km and about 30 Q-5 aircraft that could carry one bomb over 400 km have constituted the third leg of the Chinese triad. Though the first delivery capability available with China, this was

42. As cited in J. You, *The Armed Forces of China* (Canberra: Allen & Unwin, 1999), pp. 96-97.
43. For more on this, see Lin Changshneg, "The Combat Power of China's Nuclear Submarine," as quoted in Andrew S. Erickson, "China's Future Nuclear Submarine Force: Insights from Chinese Writings," *Naval War College Review*, Winter 2007.

always considered the weakest, given the short-range of aircraft and their inability to penetrate enemy air defences. But China has begun inducting H-6K bombers in an effort to improve aerial nuclear deterrence. Now fitted with D-30-P2 engines of greater thrust power and reinforced fuselage structure, the H-6K has greater range and combat payload. Armed with long-range cruise missiles modelled on the KH-55 of Russian origin with range of 2,500 km, the H-6K, even though a subsonic bomber, is now believed to have the operational capability to effectively project nuclear deterrence.[44]

Development of MIRV Technology

While it is widely assumed that China presently does not deploy multiple warheads on a single missile, it reportedly has had the capability to do so for several years. A report in December 2002 announced the test of the DF-21 medium range ballistic missile (MRBM) with multiple warheads, a capability necessary to enhance China's nuclear deterrence against BMD.[45] China is known to have been miniaturising warheads to make them lighter and easier to deploy in multiple numbers atop delivery systems. In 1999, the Cox Committee instituted to investigate allegations of Chinese espionage concluded that design information of seven most advanced thermonuclear weapons that would enable China to graduate to the next generation of smaller weapons had been stolen. However, several Congressmen, weapon designers and laboratories raised doubts about this report, and opined that even if the stolen designs were available with China, it would find it difficult to translate them into reliable warheads in the absence of testing, which stands prohibited under the Comprehensive Test

---

44. Andrei Chang, "China Attains Aerial Nuclear Strategic Strike Capability" http://www.spacewar.com, September 7, 2007.
45. "China Successfully Tests Multi-Warhead Missiles," *Yomiuri Shimbun*, February 8, 2003.

Ban Treaty (CTBT) that China has signed though not yet ratified.

Meanwhile, China has been working on developing technical counter-measures to BMD. These include the use of chaff, decoys, stealth technologies, and manoeuvring warheads. Some of these have been tested, though they are not believed to have been yet deployed. Overall, the Second Artillery Corps, tasked with the three-fold mission of deterrence, supporting conventional war with ballistic missiles attack, and nuclear counter-attack, has adopted the principle of "first resist, then penetrate," implying that Chinese nuclear forces must be able to survive a first strike and then use counter-measures to penetrate the BMD.

Rapid Enhancement of Space-Based Capabilities

While professing support for the prevention of the arms race in outer space (PAROS) in the Conference on Disarmament, it is evident that China is modernising its space-based capabilities to directly attack the enemy's space-based infrastructure with anti- satellite (ASAT) weapons, and to attack the enemy's ground-based radars or other electronic components of the command, control,communications, computers, intelligence, surveillance, reconnaisance (C4ISR) structure. Beijing has determined that a robust ASAT capability could counter the United States by disabling the "complex, exposed network of command, control, communications, and computer-based systems that provide intelligence [and] reconnaissance" to American forces.[46] In fact, it may not be a pure coincidence that the technical parameters of

46. According to a report prepared by Michael Pillsbury for the US bipartisan China Economic and Security Review Commission that is based on writings of 20 Chinese military strategists, there are three books and several dozen articles going back a decade that advocate development of ASAT to covertly shoot down satellites to send a deterrent message. As cited by Vago Muradian, "China's Mystery Satellites: US Gauges Beijing's ASAT Strategy," *Defence News*, May 2, 2007.

China's January 11, 2007 ASAT test were akin to a ballistic missile intercept. The use of a phased array radar to guide a kinetic-kill vehicle toward a target (an ageing weather satellite) travelling at 7.42 km a second, or as fast as an ICBM would be when entering the earth's atmosphere, was obviously conducted to send a message. China has also developed high altitude electro-magnetic pulse (EMP) with a nuclear device capable of exploding at 40 km altitude to paralyse electronic systems of the adversary. Also, in August 2006, a Chinese ground-based laser had blinded a US reconnaissance satellite over China.

Indeed, Chinese defence planners see the networked nature of American forces as their principal vulnerability, envisage space warfare in the future, and are developing capabilities that could blind US sensory and networking organs (surveillance, reconnaissance and intelligence assets in space). And it appears to be pursuing this in a determined fashion through the development of a range of technologies, including directed energy weapons and electronic attacks through sophisticated jamming technologies.

A White Paper entitled "China's Space Activities in 2006" places China's space programme within the overall national development strategy and stresses more investments in basic research, advance planning for frontier technologies and leapfrogging development.[47] Expansion of space capabilities for reconnaissance, navigation, communications, targeting, etc are seen as equally important as counter-space, ASAT ones. As Ashley Tellis has aptly concluded, "...China's counterspace efforts are diverse, comprehensive, rapidly improving, and deadly serious—exceeding even those of the Soviet Union at its peak. Taken together, Beijing's space denial programs leave no doubt that it is determined to negate the operational advantages accruing to Washington's space-enabled

47. *China's Space Activities in 2006* (Beijing: Information Office of the State Council of PRC, October 12, 2006).

conventional military dominance."[48] He, therefore, argues that China will continue to invest in space-denial technology rather than subscribe to any arms control agreement that eliminates its chance of asymmetrically defeating superior American military power. One PLA analyst, Wang Hucheng, articulated the same, "For countries that can never win a war with the United States by using the method of tanks and planes, attacking the US space system may be an irresistible and most tempting choice."[49]

Implications for India

While Western strategists believe that the acquisition of capabilities that would qualify for "limited nuclear deterrence" are still several years away, it may be remembered that this would be with reference to the USA. In the case of India, the ongoing process is already ominous enough. As has been pointed out by Gill, Mulvenon, and Stokes, the Second Artillery more likely operates on "three levels of doctrine: credible minimal deterrence with regard to the continental United States and Russia; limited deterrence with regard to China's theater nuclear forces; and an offensively-configured, preemptive, counter-force war-fighting posture of 'active defence' or 'offensive defence' for the Second Artillery's conventional missile forces."[50] Therefore, India is already facing not a "minimum" but a "limited" nuclear deterrence with China, which clearly leads in nuclear and missile capability, and could theoretically undertake "limited nuclear war-fighting" with counter-force targeting to impose intra-war deterrence.

However, what is interesting is that China and India have very similar nuclear doctrines. India too perceives a political role

48. Ashley Tellis, "Punching the US Military's 'Soft Ribs': China's Antisatellite Weapons Test in Strategic Perspective," *Carnegie Policy Brief*, no. 51, June 2007.
49. Ibid.
50. Mulvenon, n.24, p. 549.

for its nuclear weapons, strives for minimum credible nuclear deterrence and has declared an NFU that emphasises retaliation to cause unacceptable damage with the help of survivable and reliable nuclear forces. Given that both countries have a more stabilising nuclear doctrine than any of the other NWS, it should be assumed that there is less probability of deterrence breakdown in the case of India and China. Indeed, Sino-Indian nuclear relations are considered far more stable than Indo-Pak nuclear deterrence. However, two factors complicate the relationship. Firstly, China still does not officially recognise India as a nuclear weapon state and, hence, has refused to undertake any bilateral nuclear arms control or nuclear stability negotiations. Secondly, China's conditional NFU is believed not to apply to territories that it claims as its own (which would include Aksai Chin and nearly 90,000 sq km of Arunachal Pradesh). Hence, China's NFU can provide little assurance to India.

Does that imply that China would not hesitate to use nuclear weapons in the region in case of a crisis with India? Could the "limited use" of nuclear weapons against military troops be an option? What would Beijing gain by doing so, especially if it inflicts nuclear annihilation on the region it claims and wants to annex? Adding a nuclear wasteland would not help. Or, would China consider nuclear use against deeper Indian targets to coerce India to leave the region or to, yet again, "teach India a lesson"? Would China not be deterred by the fact that it does not yet have a disarming capability with a first strike, and its action could risk a nuclear hit on some of its own cities in an Indian retaliation that would be certain? Would "reunification" then be worth the effort? Would China be willing to risk a wider war with India? Would its modernised and informationalised strategic capabilities give it the confidence to keep the border war "limited and local"?

There can be no certain answers to these questions. In fact, arguments can be made both ways. For instance, on the one hand, it could be said that China might risk a nuclear attack even at the cost of losing one or more of its own cities, for three reasons: one, China would certainly have noticed that India does not yet have the capability to target its centres of gravity. Some peripheral cities might be lost, but then the PLA is known to have once said in the context of a nuclear war with the USA, that it was ready for the loss of cities beyond Xian; secondly, strategic modernisation would provide Beijing with confidence that it could engage in an extent of nuclear war-fighting with India without suffering "unacceptable damage to itself", by exercising escalation dominance and bringing about intra-war deterrence; and thirdly, China would do so if it found it necessary to sustain its great power image. International status and prestige is very dear to the country and its leadership and it would not hesitate to reclaim what it considers is its own in order to teach a lesson to others.

On the other hand, it could also be argued that a more developed and globally integrated China should be easier to deter. Unlike Mao who believed that his country was large and populous enough to survive a nuclear war and still carry on building Communism worldwide, modern China would have a weaker damage threshold and may not want to risk even a single city to Indian nuclear retaliation; secondly, despite how it is perceived outside, the Chinese leadership is not a monolith and would be punctuated with dissenting voices on nuclear use; and, thirdly, given the present strong moral taboo against nuclear use, China may be deterred by fear of international outrage and moral repugnance against itself. Loss of international stature as a responsible, mature international player would be a concern. But then again, there can be no certainty whether such considerations could cast enough of a restraining influence on a China that has 'risen'.

Two other factors need to be considered by India: the first is the possibility of accidental or unauthorised war with China. Though both have their nuclear arsenals in a low alert mode, once China acquires and deploys SSBNs, it would also have to put operational procedures in place. Communication problems with submerged vessels are well known and it remains to be seen how a centralised command and control system like China's deals with practical problems of delegation. Secondly, India will have to devise its responses to the deployment of some level of BMD by China itself. While this aspect is dealt with in detail in the chapter on BMD, it may be said that India would have to consider its own missile defence capabilities, at least point/area defence, especially if one believes that China would shift to a counter-force targeting strategy.

Irrespective of China's likely behaviour, the strategies for India should nevertheless be fairly clear. While, on the one hand, it must undertake conventional modernisation to raise the nuclear threshold, on the other, it needs to enhance the credibility of nuclear deterrence through a sustained programme aimed at increasing the survivability of its nuclear forces. Just as China sows doubt in the US mind that its nuclear forces are invulnerable to a disarming first strike, and willing and capable of mounting a punishing retaliation,[51] India too has to credibly convey that a nuclear attack by China would result in assured retaliation to cause unacceptable damage. Simultaneously, further improvements in the range and reliability of missiles, redundancy in the command

51. The US is perturbed by the Chinese opacity on nuclear matters. Dennis Wilder, senior director for East Asian Affaris on the US National Security Council, said in January 2009, "Chinese nuclear strategy and doctrine is really a black box." This, according to the US, impedes its ability to understand when Chinese military movements might be threatening, or, for example, how to judge various levels of Chinese military alert in order to avoid miscalculations. Robert Burns, "Bush Official Urges China to Lift Nuclear Secrecy," *The Seattle Times*, January 16, 2009.

and control systems and a clear articulation of a chain of command and succession are imperative. In fact, ironically enough, India has to follow pretty much the same path as China does towards enhancing the credibility of its nuclear deterrence.

In conclusion, the primary strategic goal of China is to build and sustain comprehensive national power (CNP). This requires internal and external stability for continued economic reforms and rise to regional preeminence. Within this scheme of things, nuclear deterrence is one component of CNP, as also a tool to attain more CNP. Therefore, China will build a small but modern nuclear force that enforces deterrence by display of its war-fighting capability. There is little likelihood of the country engaging in arms control or reduction negotiations with other nuclear powers (not that there appears any such possibility amongst the NWS), since it will not surrender a capability that it believes has brought the country stature and security. Rather, an emphasis on improving survivability would drive modernisation, though these efforts would be measured and selective. China would maintain a defensive orientation of its overall nuclear policy while, at the same time, preserving flexibility of offensive action. As it gains greater confidence in its economic strength, military capabilities and concomitant international clout, and increases its economic and political integration into the international system, its domestic, foreign and strategic policies will be publicly articulated with greater frequency and transparency. Beijing would also increase its involvement in multilateral diplomacy, not only to showcase itself as a responsible and able international player, but also to influence the security environment to better safeguard its national interests.

4

Trumpet of the Elephant
India's Nuclear Doctrine and Strategy

Having examined the nuclear doctrines and strategies of the five major nuclear weapon states, and of Pakistan, this chapter identifies and describes the salient features of India's nuclear doctrine. Unlike the case of its two neighbours, India has articulated a written nuclear doctrine and the chapter begins by explaining the critical *importance of having a clear nuclear doctrine as the basis for credible deterrence*, since in its absence, a nation's nuclear strategy could meander aimlessly and lead to misperceptions and misconstructions.[1] Irrespective of whether a nuclear doctrine professes an aggressive nuclear posture or is defensive in nature, its enunciation enables some sort of threshold identification and provides some stability to a nuclearised environment. The second section of the chapter takes an in-depth view of the *principles on which India's nuclear doctrine* as put forth by the National Security Advisory Board

1. The US Administration has been trying to get China to clarify its intentions in strategic build-up. Articulating a written doctrine could be one way of doing so.

(NSAB) in 1999[2] and subsequently in the statement issued by the Cabinet Committee on Security (CCS) on January 4, 2003[3] is based. It also *assesses the appropriateness* of these principles for India.

The Purpose of a Nuclear Doctrine

Stated simply, the purpose of a nuclear doctrine is to provide the *raison d'etre* of the nuclear weapon for a nation, as also to make available the philosophy behind fundamental questions of when, how, and where the weapon would be used for national defence. These objectives may be met through the formal enunciation of a doctrine or through hints dropped in more amorphous statements. In fact, nuclear doctrines of not all countries exist in the form of structured, publicly shared documents, as in India. As brought out earlier, in some cases, such as Pakistan, the doctrine can only be gleaned from official statements or articles written by retired military officers or political leaders or members of the strategic community.

In either case, though, the nuclear doctrine is expected to serve a critical, but only limited purpose. Limited, because there are several things that a doctrine is not. It is not a statement of account that calculates the cost of the nuclear weapons. Neither is it a fact sheet that identifies all threats and challenges faced by a country. Nor is it an operational strategy or a set of tactical rules that describes force postures or deployment patterns. Instead, the doctrine is a system of beliefs that:

---

2. National Security Advisory Board, *Draft Indian Nuclear Doctrine*, August 17, 1999. Available at http://www.meadev.nic.in. Reproduced as Appendix 1 of this book.
3. Cabinet Committee on Security, Government of India, "India's Statement on 'Operationalisation' of Nuclear Doctrine", January 4, 2003. Available at http://www.meadev.nic.in. Reproduced as Appendix 2 of this book.

(a) describes the utility of nuclear weapons to the state. It, thereby, reflects the worldview as seen by the state and the purpose served by the acquisition of nuclear weapons;
(b) identifies the manner in which the weapons would be employed to meet the purpose they have been acquired for. In performing this function, the doctrine provides direction on important subsidiary issues pertaining to force posture, concept of operations and weapon deployment.

Hence, the doctrine encapsulates the philosophy that is expected to guide national nuclear strategy. Based on these tenets, it should become possible for the national leadership, civilian and military, to determine the nature and size of the nuclear arsenal, including delivery vehicles, the kind of command and control systems, the type of retaliation and the identification of targets, deployment status, etc. Similarly, it should provide some inkling to the adversary as to how a nation intends to use its nuclear capability, at what stage of conflict nuclear weapons could come into play, and how they may be deployed.

While a doctrine must be rooted in practical realities confronting a nation, such as its strategic culture, its threat perceptions, its economic strengths and the state of existing scientific technological capabilities, it can also afford to take little flights of fancy in aspiring for an ideal situation. For instance, India's nuclear doctrine, despite seeking to provide the basis for operationalisation of nuclear deterrence, still contends,"India shall continue its efforts to achieve the goal of a nuclear weapon-free world at an early date." It identifies "global, verifiable and non-discriminatory nuclear disarmament" as a "national security objective."[4] Similarly, the doctrine also establishes the need for India's nuclear forces to be "based on a triad of aircraft, mobile

4. As stated in Para 8.1 of the *Draft Indian Nuclear Doctrine*.

land-based missiles and sea-based assets".[5] While neither of these situations is immediately available, the doctrine encourages movement towards them in the interest of national security. Therefore, unlike a nuclear strategy that must be solely dictated by existing realities and capabilities, the doctrine can display a greater sense of freedom and flexibility.

A coherent document on India's nuclear doctrine was formulated just fifteen months after the conduct of nuclear tests in May 1998. In one of its earliest tasks, the first NSAB[6] constituted soon after India declared itself a state with nuclear weapons, produced a draft that encapsulated recommendations on what the country's nuclear doctrine should be. On August 17, 1999, the body presented the draft nuclear doctrine to the National Security Council and through it to the government, which immediately made it available for public scrutiny and debate. The national and international community were likewise taken by surprise not only by the speed with which the task was undertaken, but also with the transparency that the then caretaker Indian government offered on a subject that in most countries is normally considered elitist and kept out of public purview. Not surprisingly, therefore, the Indian doctrine was subjected to a fair amount of criticism from within the country and outside, not all of which was either justified or constructive.

Amongst the responses from the international community, the US, in its capacity as the self-appointed spokesperson of global non-proliferation, promptly expressed its disapproval of the draft doctrine. The White House did not find the draft an "encouraging

5. As stated in Para 3.1 of the *Draft Indian Nuclear Doctrine*.
6. The NSAB exists as an official body that is part of the National Security Council. It acts as a forum through which the government decision-making apparatus can draw on the advice and experience of appointed academics and retired civil servants and military officers. Members of the NSAB serve a term of two years.

document" and the Pentagon too pointed at the dangers inherent in such an act and called for the denuclearisation of South Asia. Russia and China too expressed concern at the Indian move and urged restraint. Perhaps it was felt that the doctrine granted a legitimacy and sense of permanence to India's nuclear weapons.

While the P-5 were shocked at the audacity of a newly nuclear country to be able to articulate a nuclear doctrine in such a short time after nuclearisation (considering that the US took 15 years to formulate its nuclear doctrine), there seemed to be some amount of confusion in Pakistan on how much importance it should attach to the Indian action. While, on the one hand, Pakistan criticised India's move in bringing out the document, as also the substance contained therein for being "dangerous and ominous", on the other, in a typical "me too" response, it quickly announced that Islamabad too was engaged in giving "final touches" to its own doctrine![7] While no doctrine has since been officially announced by Islamabad, a statement attributed to the then Pakistani Foreign Minister revealed what Pakistan actually thought about the nuclear doctrine. He was quoted in *Dawn* as saying, "By announcing its nuclear doctrine before Pakistan, India was *trying to score points and present itself as a more responsible nuclear power in the region.*"[8]

As derived from here, one would assume that the release of a nuclear doctrine actually amounts to a display of responsible behaviour. This, in fact, is the truth. The declaration of a nuclear doctrine by a country that has unambiguously demonstrated its nuclear capability actually clears the air and injects transparency into an issue otherwise shrouded in secrecy. By clearly defining a political role for its nuclear weapon and the philosophy behind

7. "Pakistan Reacts Strongly to India's Assertion," *The Times of India*, August 19, 1999.
8. M. Ziauddin, "Pakistan's Nuclear Doctrine being Finalised: FM," *Dawn*, August 19, 1999 (emphasis added).

its use, New Delhi provided a rare clarity. While this was, not unexpectedly, construed as India's march towards nuclear build-up, the doctrine actually only formalised what India had declared as its nuclear position all along after the nuclear tests in May 1998. The draft doctrine reflected the same precepts that had been articulated by then Prime Minister Vajpayee in Parliament within days of the tests.[9] It emphasised the acquisition of a *credible minimum deterrence (CMD)* and advocated a *no-first use (NFU) posture* premised on a *counter-strike capability to inflict unacceptable damage.* With each one of these parameters, India had chosen to impose restraints and checks upon its own self.

In fact, it is this restraint that was criticised by several quarters of the Indian strategic community. While a few decried the doctrine for being too ambitious since it spoke of a triad[10], there were others who objected to the concept of minimum deterrence[11], and yet others who dismissed minimum deterrence as "making a virtue out of necessity."[12]

Perhaps, it should be conceded that India's choice of

---

9. Suo Motu Statement by Prime Minister Vajpayee in the Parliament on May 27, 1998. As reproduced in *Strategic Digest,* July 1998.
10. Praful Bidwai, "New Delhi's Nuclear Doctrine Could Ignite Arms Race," *Dawn,* August 23, 1999.
11. Bharat Karnad, for example, advocated a "maximally strategic" deterrence posture built around multiple kinds of high yield nuclear weapons and many delivery systems. For more on his views, see Bharat Karnad, "A Thermonuclear Deterrent," in Amitabh Mattoo, ed., *India's Nuclear Deterrent* (New Delhi: Har-Anand Publications,1999), p. 108.
12. Brahma Chellaney wrote, "India does not have the plutonium or financial resources to exercise more than the barest of minimum deterrence, and is far from having the capacity to carry out a disabling first strike against an opponent..." "India's Nuclear Planning, Force Structure, Doctrine and Arms Control Posture," paper presented at the Forum of the UNESCO International School of Science for Peace on Nuclear Disarmament, Safe Disposal of Nuclear Materials, or New Weapons Developments? Where are the National Laboratories Going?, Landau Network-Centro Volta, Cernobbio-Como, Italy, July 2-4, 1998.

minimum deterrence is to some extent dictated by the resource (not just monetary, but also available fissile material) and technological (missile range, accuracy and payload carrying capability) constraints that the country faces, but as is argued in the body of this chapter, CMD more than adequately addresses India's security concerns that are required to be met with nuclear weapons.

Its criticism notwithstanding, to its credit, the draft report of the first NSAB on the Indian nuclear doctrine provided the essential starting point for a coherent nuclear strategy. On January 4, 2003, the government issued a statement on the basis of a decision taken by its Cabinet Committee on Security (CCS), which further amplified the nuclear doctrine and operational arrangements governing India's nuclear assets. While this 'official' nuclear doctrine largely retained most of the recommendations of the NSAB document, it did, however, make a few changes, not all of which have been particularly helpful in explaining the nuances of deterrence. Some, in fact, such as the expansion of the role of nuclear weapons to deter chemical and biological weapons, as also the use of the term "massive retaliation" in place of "punitive retaliation" have created some confusion. These are discussed in greater detail in the following sections.

India's Nuclear Doctrine – The Underlying Principles

Use of Nuclear Weapons for Deterrence

From the time of the first demonstration of the destructive power of the nuclear weapon in 1945, there has been considerable soul searching on the actual role of nuclear weapons in national security. In 1950, George Kennan pithily articulated this crucial question before the then US Secretary of State:

> Are we to rely upon weapons of mass destruction as an integral and vitally important component of our military strength, which we would expect to employ deliberately, immediately, and unhesitatingly in the event that we become involved in a military conflict with the Soviet Union? Or are we to retain such weapons in our national arsenal only as a deterrent to the use of similar weapons against ourselves and as a possible means of retaliation in case they are used?[13]

This dilemma – whether the nuclear weapon is just another type of weapon that is militarily usable or whether the absolutely horrific consequences of its use place it in the realm of the unusable except as a political instrument of deterrence – has preoccupied every country that possesses it. India has resolved the issue for itself by accepting the latter use of the nuclear weapon. Accordingly, the Indian nuclear doctrine is firmly rooted in the belief that nuclear weapons are a political instrument for deterrence and not a military tool for war-fighting. This assertion, for India, emanates from two facts: one, from the comprehension of, and abhorrence for, the high destruction potential of the nuclear weapon that makes its use unthinkable for any rational political end. There can be no cause worthwhile enough to justify the actual use of nuclear weapon; second, from the realisation that there is no credible defence against nuclear weapons. Way back in the mid-1950s, a British scientist, P.M.S. Blackett had conceded, "There is no effective defence at present, nor is there one in sight, against a *large-scale and determined atomic attack on cities and centres of population.*"[14]

13. As cited in Lawrence Freedman, *The Evolution of Nuclear Strategy*, Third Edition (New York: Palgrave Macmillan, 2003), p. 63.
14. P.M.S. Blackett, *Atomic Weapons and East-West Relations* (Cambridge: Cambridge University Press, 1956), p. 3. It may be underlined that BMD is not nuclear defence but defence, not guaranteed either, against one type of a nuclear delivery vehicle.

Hence, the Indian nuclear doctrine, with its emphasis on deterrence, actually seeks to obviate the possibility of the use of the nuclear weapon in the first place. It aims to persuade the adversary that any nuclear use by him would result in such retaliation as to make the cost of action unacceptable. It compels a reconsideration of the decision to use nuclear weapons at all.

The second important characteristic of India's nuclear doctrine, at least as put forth in the NSAB draft, was to restrict the possible use of nuclear weapons only against the adversary's use of nuclear weapons. In other words, India perceived a role for its nuclear deterrence only against the use or threat of use of other nuclear weapons and not against conventional (as NATO's and Pakistan's nuclear doctrines profess) or other WMD such as chemical or biological weapons (as the US doctrine contends). Of course, there were analysts who considered this view too restrictive. For instance, Brahma Chellaney argued,

> It should not be simplistically assumed that nuclear weapons are only for deterring a nuclear threat, as if it is okay for others to employ conventional force against vital Indian interests. Given the decades-old technology sanctions it has suffered and the way its foreign policy has been seriously constricted, India has had to pay an extraordinarily heavy political price for its nuclear weapons. It should seek to derive the full security value from them, the way the wealthy nuclear powers do.[15]

Perhaps, this rationale ultimately did hold sway over nuclear decision-makers since, in the January 2003 official statement, the scope of nuclear deterrence was expanded to include chemical and biological weapons too. It may be recalled that the US too had done the same in its Nuclear Posture Review of 2001. The move, however, made little sense for the US and

15. Chellaney, n.12.

hardly makes the Indian nuclear deterrent more credible. This is for two reasons: first, the international community has already outlawed the use of these two classes of WMD through the Chemical Weapons Convention and the Biological and Toxic Weapons Convention respectively. Countries signatory to the two conventions are supposed to have declared and destroyed their stockpiles. Therefore, India's potential adversaries should not normally be assumed to possess these weapons. In case India still fears such an eventuality, it can seek redressal by invoking its right to inspections. Despite these, if there are still apprehensions about any clandestine stocks of the adversary being used in conflict, even then, should India retaliate with its nuclear weapons? Would this then not escalate the situation into a sure nuclear exchange? What then should India's response be? While it is beyond the scope of this chapter to discuss this scenario at length, it may be said that conventional capabilities should be able to, or made able to, handle such a situation.

Second, the contemporary threat perceptions, in fact, accrue greater probability to these WMD being used by non-state actors for whom treaties are irrelevant. But against them, classical nuclear deterrence can anyway not hold good. Non-state actors do not provide a fixed target that can be deterred with unacceptable damage through the use of nuclear weapons against them. A different cache of military and foreign policy tools needs to be employed against them. Therefore, except for some symbolic value, the expansion in scope of the nuclear doctrine holds little relevance for nuclear deterrence as required by India against its potential adversaries.

Instead, India's narrow articulation of nuclear weapons as a means of deterrence only against nuclear weapons actually accurately reflects its traditional abhorrence for nuclear weapons and the reluctant steps it took down this path. It underscores that India had consistently argued that nuclear disarmament, and not nuclear deterrence, can and must constitute the basis for lasting world peace and security. But the absence of any substantive progress on this front coupled with a deteriorating regional security environment compelled India to acquire the nuclear weapon. Yet, it continues to view it as a pure deterrent. As Prime Minister Vajpayee said soon after India's nuclear tests, New Delhi does not "intend to use these weapons for aggression or for mounting threats against any country; these are weapons of self-defence to ensure that India is not subjected to nuclear threats or coercion."[16]

Credible Minimum Deterrence

The rejection of the concept of nuclear war-fighting frees India from the need to match the nuclear arsenal of its adversary/(ies) weapon for weapon. As Kenneth Waltz stated several decades ago, "Forces designed for war-fighting have to be compared with each other. Forces designed for war-deterring need not be compared. The question is not whether one country has less than another, but whether it can do unacceptable damage to another...."[17] With the principal role of India's nuclear force being to protect the nation from nuclear blackmail and coercion, instead of any desire to enforce compellence, or mount aggression, the country's policy-makers perceive the need for only "credible minimum deterrence". Official pronouncements, however, have refused to be drawn into quantifying the minimum deterrent. Rather, the draft doctrine leaves the decision on the actual size and composition of the nuclear arsenal to threat perceptions from the strategic environment, the technological imperatives and the needs of

16. Suo Motu Statement, n.9.
17. Kenneth Waltz as quoted by Gen Sundarji, *The Blind Men of Hindoostan* (New Delhi: UBSPD Ltd., 1993), p. 68.

national security. As Jaswant Singh, India's Foreign Minister in 1998, said,

> The minimum is not a fixed physical quantification. It is a policy approach dictated by, and determined in, the context of our security environment. There is no fixity. Therefore, as our security environment changes and alters and as new demands begin to be placed on it, our requirements too are bound to be evaluated.[18]

While the determination of CMD would change with transformation of threat perceptions and technological developments, it definitely need not seek superiority or even parity with an adversary's nuclear forces in the numbers,yields or types of weapons. However, this freedom is qualified by the need to acquire an assured capability of a counter-strike that can inflict unacceptable damage on the enemy. CMD based on such a principle, therefore, imposes its own prerequisites, as defined below:

- Sufficient, survivable and operationally "ready" nuclear forces. This implies that while mere numbers and balance of force might not be relevant because the scale of destruction caused by even a few nuclear weapons could constitute unacceptable damage, what is critical is to ensure that *sufficient warheads and delivery vehicles survive* a first strike and can be *ready for retaliation*. This calls for the assured survival of a requisite retaliatory strike potential and, hence, the need for survivability measures such as hardened silos, mobile launchers, deployment beyond the reach of hostile delivery systems, dispersion of the arsenal on a triad, and a pre-determined chain of succession for the weapon release authority. Reliability of the delivery system is critical for

18. Foreign Minister's speech in Parliament on December 16, 1998. Downloaded from http://www.meadev.gov.in.

deterrence credibility and this includes dependability of communication (that the correct message is delivered at the right time for launch); of launch (that the missile actually lifts off); of the booster (that it ignites in time); of separation (of the booster from the missile after burn out); of penetration (despite enemy air defence systems); and of detonation (at the designated target).[19]

- Robust command, control, communication and intelligence systems (C3I). This is critical to ensure that the nuclear assets *remain secure in peace-time* but can make the *shift to fully employable forces* when necessary in the shortest possible time for effective retaliation. Such a system comprises personnel, procedures that acquire, collate, analyse and interpret information to assist decision-making, and equipment that enables acquisition and transmission of decisions to different constituents of the force in real-time.
- Effective surveillance and early warning capabilities to acquire intelligence on alert status of the adversary. These are necessary to *minimise risks of a miscalculated or inadvertent strike* based on faulty intelligence or false alarms.
- Comprehensive planning and training for nuclear operations consistent with strategy. *Well-trained and motivated human resources* are the key to effective operations, especially when dealing with nuclear weapons. The soldier must be thoroughly prepared, technically and psychologically, to handle the responsibility.
- The "political will" to employ such forces if and when required. The actual use of the nuclear weapon is ultimately a political decision. Therefore, for deterrence to be credible *visibility of political will* through an organisational set-up reflecting institutional decision-making is crucial. There is dire need

19. For more on this, see Sundarji, n. 17, p. 76.

for conducting periodic scenario building exercises and regular threat assessments to enhance the understanding of the political leadership of the sophisticated game of nuclear deterrence and the value of correct nuclear signalling by the right authority.

No First Use Against NWS and Non-Use Against NNWS

The central principle that logically flows out of the designation of nuclear weapons as political instruments of deterrence is their no first use in a conflict. Doctrines that ascribe a war-fighting role to nuclear weapons need to adopt aggressive postures that envisage their first use. During the Cold War, the USA and the USSR believed that a nuclear war could be fought and won and, hence, went on adding numbers and newer delivery capabilities in order to maintain an edge over the other. For the Americans, crafted as their war strategy was on the Pearl Harbour experience, acting first and maintaining surprise were critical. Not surprisingly, therefore, Washington subscribed to a LOW and LUA postures. This was done in the belief that unless the US was able to undertake a preemptive/surprise strike, it stood little chance of being able to destroy all Soviet targets as required in its war plan. And minimising, if not completely eliminating, the enemy's second strike capability was the primary task of the first strike, hence, the need for ever increasing numbers of warheads. NATO too adopted a first use doctrine to deter Soviet conventional might, a logic that Pakistan now uses in support of its nuclear doctrine of first use against India. The Soviet Union, similarly, relied on its capacity to mount preemptive attacks. And, to undertake preemption, both sides built up large, ready, counter-force capabilities supported by a huge infrastructure in the form of C3I, early warning systems, etc.

India's nuclear doctrine, in contrast, has freed itself of many of these requirements by basing its nuclear strategy on a *retaliation only* policy.[20] It implies that India will not be the first to introduce nuclear weapons in any conflict with a nuclear weapon state (NWS) and not use nuclear weapons in a conflict with a non-nuclear weapon state (NNWS), unless it is aligned with an NWS. Until now, the only nations to have publicly adopted an NFU have been the USSR and China. The former had formally pledged itself to NFU in 1982 but abandoned it a decade later on the Soviet disintegration in 1992. China had announced an NFU in 1964, but has since made it conditional. In recent times, at least three conditions have been specified for first nuclear use: one, an attack on Chinese ICBMs; second, an attack on Chinese strategic hubs, such as Shanghai and other industrial and commercial centres; and,third, territories claimed by China as its own.[21] Therefore, India's unconditional NFU has no precedent and understandably, is greeted with scepticism.

Within the country itself, there are enough who argue against no first use on the ground that it would jeopardise national security. Some contend that it would lead to a need for building a larger nuclear arsenal than would be required with a first use policy. Others dismiss the NFU doctrine as nothing more than public posturing not capable of offering any guarantee against first use if the need so arose. While this is true of any declaratory policy, the fact remains that a no first use doctrine makes sense once it

---

20. As early as 1985, India had decided that NFU suited it best. This conclusion was reached by the small group set up in 1985 and comprising then VCOAS, K. Sundarji, VCNS, K.K. Nayyar and DCAS, John Greene along with R. Chidambram of DAE and Abdul Kalam of DRDO to work out the cost of an Indian nuclear deterrent. See George Perkovich, *India's Nuclear Bomb: The Impcat on Global Proliferation* (Berkeley: University of California Press, 1999), pp. 273-274.
21. As explained by Dr Srikanth Kondapalli, Professor, School of International Studies, JNU, New Delhi, and noted Indian China expert, in a private conversation with the author.

gets translated into force postures, and in that manner, provides a measure of indication of a country's intentions. While a first use posture requires missiles to be on alert for LOW/LUA and the nuclear warheads to be mated or ready to be instantly mated with the delivery system, NFU offers greater time and a more relaxed posture. In fact, several points of criticism of NFU do not stand up to scrutiny. If analysed dispassionately, as has been attempted in the following paragraphs, NFU offers the best possible choice for India in the present circumstances.

First of all, it must be understood that the adoption of NFU does not in any way adversely impact India's ability to defend itself against nuclear weapons. Given that the country does not foresee any plausible, rational scenario for the actual use of nuclear weapons, and least of all where it might be compelled to use nuclear weapons first, not for coercion and nor for any territorial expansionist ambitions, NFU appears most logical. By placing the onus of escalation on the adversary, while retaining the initiative of punitive nuclear retaliation, India has sought to steer away from nuclear brinkmanship in any and every conflict. Meanwhile, by establishing the nuclear weapon as an instrument of punishment, India seeks to prevent deterrence from breaking down, and, thus, aims to minimise, if not prevent, the very use of the nuclear weapon. NFU actually encourages the possibility of "no use" instead of "sure use". This is clearly demonstrated in Table 4.1.

Table 4.1

Nuclear Posture Country 1	**Nuclear Posture Country 2**	**Nuclear Threshold**	**Chance of nuclear war**
First Use	First Use	Low	Very High
No First Use	No First Use	High	Very Low/Nil
First Use	No First Use	Relatively High	Low

There are some who argue that by binding itself to an NFU pledge, India has left the initiative with the adversary. But as is evident from table, through NFU, coupled with assured retaliation, India has reined in the initiative more in favour of no use of nuclear weapons. Unless the adversary's leadership is completely irrational, has suicidal tendencies or is utterly unmindful of national survival and international public opinion, the possibility of a nuclear war should not arise and the surety that India would not use nuclear weapons first would ensure that.

Secondly, it is also questioned whether India should retain NFU even if it gets to know that the adversary is preparing for a nuclear strike? Should not preemption then be the right step? The answer to this lies in understanding that even preparation is no guarantee of a nuclear strike. Rather, it may well be part of a strategy of "coercive diplomacy". It is not a coincidence that all the 51 incidents of threat of use of nuclear weapons actually were intended to conduct coercive diplomacy. Therefore, despite the apparent show of readiness, there will, more likely than not, still be a chance that nuclear weapons would not actually come into use. But by striking first, India would end up inviting certain and massive retaliation. In fact, even if the adversary's first use of nuclear weapons was planned as a small demonstration strike, with India's preemptive nuclear strike, it would surely retaliate with all that is available. In such a scenario, therefore, it would be better to indicate own level of nuclear preparedness in order to reinforce deterrence, seek international diplomacy to mount pressure, and, at the same time, maintain readiness of credible and nuclear conventional forces.

In fact, NFU is the best answer to those international strategic analysts who believe that nuclear weapons in India and Pakistan lead to a condition of instability.[22] They argue that since both

22. This has been forcefully argued by Sumit Ganguly and Kent L. Bringer, "Nuclear Crisis Stability in South Asia," in Lowell Dittmer, ed., *South Asia's Nuclear Security Dilemma : India, Pakistan and China* (New York: M.E. Sharpe, Inc., 2005), pp. 29-48.

sides have small nuclear forces, they would be tempted to launch a disarming (to the extent possible) first strike in case of a crisis. But India's no first use posture removes this temptation not just for itself, but also for the adversary. Secondly, NFU necessitates measures for increased survivability in order to reduce the vulnerability of the nuclear arsenal and these too mitigate the 'use or lose' syndrome. In fact, NFU goes to alleviate Pakistani insecurity, which, in turn, is beneficial to India. It relieves pressure on Pakistan's leaders for launching a preemptive strike. If Pakistan were constantly under the fear that an Indian nuclear strike was imminent, its temptation to use its nuclear force would be higher. Therefore, for the sake of crisis stability, it is actually in the best interest of India to make its adversary feel more secure, rather than on the defensive and mistrustful of Indian nuclear intentions. This situation was best described by Robert McNamara in the context of the Soviet hardening of their missile sites. He wrote,

> In a period of tension, I wanted the Soviet leaders to have confidence that those forces would survive an American attack and would be capable of retaliating effectively. Then they would not feel a pressure to use them preemptively... I had no desire to face, in a period of tension, an adversary who felt cornered, panicky and desperate and who might be tempted to move irrationally.[23]

Thirdly, by declining a first use stance, India has also removed the need for retaining nuclear forces on hair trigger alert, a situation not at all conducive to strategic stability given the geographical realities of the neighbourhood. Having nuclear forces on alert not only raises the possibility of an accidental nuclear war based on a gross miscalculation, but also lowers the threshold of nuclear war

23. Robert McNamara, *Blundering into Disaster: Surviving the First Century of the Nuclear Age* (London: Bloomsbury, 1987).

in a crisis situation. In the case of India and Pakistan, this would be akin to inviting trouble, given their proximity, low warning time and frequent crises. Therefore, India's NFU actually brings stability to the nuclear equation. It allows for a policy of recessed deterrence that allows nuclear weapons and delivery vehicles to be developed and built, but to be stored separately, ready to be assembled in the event of a crisis.

It may be recalled that in the early years of the Cold War, superpower warheads were not routinely mated, nor necessarily co-located, with delivery systems. It was the subsequent development of safety features designed into modern warheads and the advent of sophisticated administrative controls on nuclear weapons that made higher alert levels possible. Ironically, however, after keeping their nuclear missiles on hair trigger alert for years, the two superpowers found the best nuclear risk reduction measures in de-alerting these and separating warheads from delivery systems! De-mating, de-alerting and de-targeting, the three steps taken by the superpowers for nuclear risk reduction and confidence building form a natural part of the NFU posture.

Without having to go through this cycle, India's nuclear doctrine accepts the adoption of NFU as more stabilising since nuclear forces need not be maintained on high alert status. In fact, for deterrence to be credible with a no first use doctrine, it would only be necessary to have available all the relevant nuclear assets, though dispersed, as unassembled nuclear warheads under civilian control, and dedicated delivery systems kept either in storage or in readiness away from their operational areas—as long as they can be brought together as rapidly as required during a supreme emergency.

Fourthly, NFU also answers another argument made against nuclear weapons causing regional instability. It has been alleged that since the arsenals of India and Pakistan are small and technically primitive, they lack fail-safe devices such as double keys, permission action links (PAL) or other elaborate procedures insuring against an unintended or accidental attack. NFU eases this situation since it precludes the need for delegation of authority for launch and, thus, minimises the risks of miscalculation and accidental attack. In fact, given that the existent non-proliferation constraints limit the possibility of such devices becoming available to India, NFU makes the task easier.

One situation, however, that could test India's NFU is a scenario where a Taliban type military man or some non-state actor (NSA) takes control of nuclear weapons in Pakistan after a period of political instability.[24] Such a scenario could, however, be viewed from two perspectives. Firstly, if the new player has assumed state power and become a political actor, then he would also develop a stake in the political system and powerdom and, hence, would be expected to display some sense of rationality in use of nuclear weapons. Classical deterrence based on a retaliation policy to cause unacceptable damage can be expected to apply in such a scenario. But in case the NSA has got hold of nuclear weapons in a situation of political chaos, and threatens to use it against India in a suicide bomber mentality – in order to wreak nuclear havoc without worrying about the consequences of the same for his own state – then nuclear deterrence becomes more difficult to apply. To deal with cases such as this, India needs a multi-pronged strategy: one, it must maintain high conventional capability that can be used preemptively in order to stop a nuclear attack on itself; second, it must declare that WMD terrorism would invoke

24. The prospect of a nuclear capable state losing control of some of its weapons to terrorists has been put forth as a real and immediate danger in the US Quadrennial Defense Review 2006. American options in case of such an eventuality have been well brought out in Thomas Donnelly, "Choosing Among Bad Options: The Pakistani 'Loose Nukes' Conundrum," *National Security Outlook*, American Enterprise Institute for Public Policy Research, May 2006.

retaliation, of whatever nature may be necessary, against the state known to be sponsoring such activities; third, it must participate in global/multinational endeavours aiming at controlling proliferation of dual use materials through strengthened export controls, and enhanced security and safety of nuclear arsenals; and fourth, it must focus on better intelligence and preparedness levels to mitigate a national disaster.

Evidently then, no first use can hold in a range of situations, and in fact, the posture that it translates into, offers several advantages for India:

(a) It eliminates the need for forward deployment of nuclear systems, and, thus, reduces the likelihood of accidental or unauthorised use. LOW/LUA force postures, by their very nature, require pre-delegation of authority to launch nuclear weapons down a clearly defined chain of command. It may be mentioned in this context that the US and USSR, which for many years relied on tactical nuclear weapons (TNWs), were never unaware of the dangers inherent in these postures. It was found that in such a situation, the "fog of war" was abnormally dense. Battles where use of TNWs was envisaged were described as "battles of great confusion; [where] casualties would be high; troops would be left isolated and leaderless; and morale would be hard to maintain."[25] Therefore, TNWs were amongst the early ones to figure in arms control initiatives once the two superpowers had realised that "the use of nuclear weapons could never be a purely 'tactical' decision, taken by the local commander according to the state of battle. It would be a strategic decision to be taken at the highest level and with reference to the prevailing, overall political and military situation."[26] India has adopted this wisdom without the experimentation.

25. An assessment of William Kaufmann as cited by Freedman, n.13, p. 104.
26. Ibid., p. 105.

(b) It forecloses the chance of an irrational preemptive strike, as also minimises the risks of an inadvertent or unauthorised nuclear use. In times of crisis, lack of information, misinformation and misjudgments can often be the cause of confrontation without either side having the intention to precipitate one. As McNamara said, "It is correct to say that no well informed, coolly rational political or military leader is likely to initiate the use of nuclear weapons. But political and military leaders, in moments of severe crisis, are likely to be neither well informed, nor coolly rational."[27]

(c) It reduces pressures on the C3I systems. Obviously, first use of nuclear weapons based on LUA/LOW places a greater strain on communications. But recessed deterrence can make do with a less sophisticated, elaborate, and, hence, less costly system.

(d) It relieves pressure from an immediate response to nuclear attack. As C. Raja Mohan has argued, "India believes its deterrence requirements can be met without time-urgent responses to a nuclear attack."[28] A similar point was made by Gen K. Sundarji, in the late 1980s, "The response can be a good few hours or even perhaps a day after the receipt of the first strike. A very highly sophisticated, highly responsive command, control and communication system that functions in real-time is not necessary...." Even though today and in the future, the timelines for retaliation will be shorter than Gen Sundarji envisaged owing to the likelihood of quickly building international pressure to abort any response, NFU does, nevertheless, provide the luxury of certainly more time than LOW/LUA postures.

---

27. McNamara, n. 23, pp. 13-14.
28. C. Raja Mohan, "No First Use and India's Nuclear Transition,",Pugwash Papers, Pugwash Meeting 279, London, November 15-17, 2002.

Besides the above-mentioned benefits, there can be no disputing the fact that a no first use policy is morally the most correct one. That is, if there can be anything morally correct about nuclear weapons at all. Nuclear weapons are not just any weapons that could or should be used indiscriminately. They are special in the sense of their immense destructive capability. And therein lies their value as deterrents.

More than anything else, the merit of India's NFU policy lies in challenging the long held nuclear theology of first use as professed by the Western nuclear powers. Until now, this has been touted as the only possible approach to use nuclear weapons for safeguarding national security. The Indian adoption of an unconditional NFU has opened up another possibility at the other end of the spectrum. It offers a counter-view to the traditional aggressive and arms race generating doctrines. If this were to be accepted by all NWS, then the world might find itself on its way to a diminishing salience of nuclear weapons, perhaps their de-legitimisation, and eventually their abolition.[29]

Punitive Retaliation

India's nuclear doctrine frees itself of the compulsion of immediate retaliation and bases deterrence instead on the *certainty* of retaliation that would be *punitive* in nature. Therefore, while the counter-attack may not be prompt, it would be assured and would cause unacceptable damage to the adversary. As has been said, "The ability to retaliate with certainty is more important than the ability to retaliate with speed."[30] In fact, the time taken to retaliate would have to be dependent on technical and practical realities such as the time required to bring together the nuclear weapon

29. For more on this subject, see chapter 11 of this book.
30. Ashley Tellis, *India's Emerging Nuclear Posture: Between Recessed Deterrent and Ready Arsenal* (Santa Monica, CA: RAND, 2001), p. 326

and delivery vehicle, the nature of precise command and control and custody arrangements, the state of the country after having absorbed the first strike, as also other domestic and international factors.

Accordingly, therefore, India's doctrine does not specify any time for retaliation. In fact, the NSAB draft did not even define the nature or magnitude of punishment, except for describing it as "punitive". Beyond this, it did not address questions about the character, extent or weight of retaliation. Instead, it followed the French logic that "the adversary must not be able to calculate what would be the reaction to this or that initiative he might take."[31] It only conveyed that retaliation would be certain and it would be devastating, irrespective of when or how it was inflicted.

However, while the draft nuclear doctrine mentioned "punitive retaliation," the 2003 official version changed it to "massive retaliation." The reasons for doing so have been conjectured upon by many.[32] But this has not necessarily enhanced the credibility of deterrence because it actually restricts the available response to an adversary's first strike to only an all out nuclear attack. This may appear too drastic for use except in extreme circumstances. While it may be argued that India's nuclear philosophy does not conceive nuclear use except in extreme circumstances, however, a case can be made for greater flexibility in response options. In fact, "punitive retaliation" is credible enough since it provides alternatives relative to the nature of strike and level of provocation. The US too in the 1960s had reached the conclusion that it should have a variety of response options other than massive retaliation against cities. This wisdom was obtained after several US officials, including Robert McNamara as Secretary of Defence, expressed their dissatisfaction

31. This is the logic of the French nuclear doctrine, as explained by Ashley Tellis. Ibid., p. 321.
32. See Rajesh Rajagopalan, *Second Strike: Arguments About Nuclear War in South Asia* (London: Penguin Books, 2005) and Ibid.

with the inflexibility of an SIOP that envisaged a preemptive first strike involving 3,423 weapons totalling 7,847 megatons against the Russians and Chinese in case of any conflict, irrespective of the provocation!![33] A more flexible response was, therefore, proposed that envisaged a substantial raising of the nuclear threshold for the critical initial responses to be made by conventional forces alone, keeping the use of nuclear weapons late and limited.

In the Indian context too, especially vis-a-vis Pakistan, as Ashley Tellis argues, by basing deterrence on massive retaliation that forecloses the option of a graduated response, India might end up encouraging its adversary to massively employ its own arsenal in the fear that India's "massive retaliation" to any nuclear use would anyway hold the possibility of disarming it. [34] Thus, India might invite a greater nuclear use upon itself than the enemy might actually have factored into its calculations. This would then rob India of the opportunity to exercise escalation dominance if it even became necessary to contemplate nuclear use. Meanwhile, with regard to China, given its existing nuclear superiority and higher survivability quotient, Indian threats of massive retaliation after having suffered a first Chinese strike hardly seem credible.

Civilian Control over Nuclear Weapons

India's nuclear doctrine establishes strict civilian control over any decision to use nuclear weapon in conflict, as also over the custody of the nuclear warheads in peace-time. This is a clear reflection of the constitutional system of governance where the ultimate decision-making is the responsibility of the Prime Minister. While the nuclear force is expected to be maintained in the form of separated components, with the responsibilities for the command,

33. For more on this, see DeGroot, *The Bomb: A History of Hell on Earth* (London: Pimlico, 2005, pp. 268-269.
34. Tellis, n. 29, p. 340.

custody, integration, and use of the weapons clearly demarcated between the civilians and the military, the command over their use would lie solely with civilian leadership. In the remote contingency that deterrence does breakdown and nuclear release orders are issued by the Prime Minister (or his designated successors), the nuclear components would be integrated into a usable weapon system, with custody to be transferred to the military, which would retain sole responsibility for executing nuclear use options.

Pursuit of Universal Nuclear Disarmament

Ordinarily, one should not expect to find a mention of nuclear disarmament in a nuclear doctrine in the first place since it actually seeks to operationalise a nuclear weapons policy. However, India's nuclear doctrine clearly designates universal nuclear disarmament as a "national security objective". It identifies a nuclear weapons free world (NWFW) as the ideal state of affairs but since the world did not seem to be getting any closer to it, the Indian nuclear doctrine was compelled to bring the Indian nuclear strategy more in alignment with the world of *realpolitik*. It may be recalled that while explaining the rationale behind India's overt acquisition of the nuclear capability to the Parliament on August 4, 1998, the then Indian Prime Minister had in the same breath, also exhorted India's nuclear diplomacy not to lose sight of the objective of an NWFW.

Counter-Value Targeting

With a no first use posture, if in the remote contingency that India is to respond to a nuclear attack, it would be most logical to use the weapons on cities instead of on purely military targets. While this sounds barbaric, it is actually the threat over large chunks of population and the productive assets of the country that can constitute unacceptable damage to make deterrence best work. Also, given that the accuracy measure of Indian missiles might be less than optimum in some ranges, Indian nuclear weapons would be most effective in attacks against economic and industrial assets, infrastructure nodes, and population centres. According to Ashley Tellis, "India's relatively small number of low-yield weapons is not optimized for effective direct attacks on opposing nuclear forces (counter-force targets)…."[35] On the other hand, they would be best suited against cities that denote softer targets (easy to pulverise even with a low yield weapon), larger targets (less sensitive to accuracy of delivery systems) and easier to locate targets (even without the help of very sophisticated navigation and targeting systems). The damage inflicted upon life and property would certainly constitute punishment of a magnitude that any country would find unacceptable.

Security and Safety of the Arsenal

The doctrine places a lot of emphasis on ensuring the security and safety of the nuclear arsenal, not just in peace-time but also in war-time. Given that India's nuclear arsenal exists in a de-mated state and in different locations, the risk of unauthorised use or the chance of an inadvertent use due to miscalculation is less but it also means having to provide for the physical security of the assets to guard against theft, sabotage or unauthorised access. It also calls for the need to match *ease of storage for ensuring safety with ease of availability at the launch site or airfield when required.*

Conclusion

Even though it must be said that a doctrine is not cast in stone since it must reflect the realities – political, economic and technological – of the times and could change as these parameters undergo a transformation, nevertheless, for the time that it exists, a nuclear doctrine performs the crucial task of providing a window to how

35. Ibid., p. 352.

a country perceives its nuclear weapons. It explains why it needs these WMD and how it plans to use them in the achievement of those objectives. India has premised its need for nuclear weapons on its desire to resist nuclear coercion or blackmail and, hence, envisages its use only for self-defence. Accordingly, New Delhi has enunciated a nuclear doctrine that perceives a purely political role of deterrence for its nuclear weapons. Flowing therefrom, India's nuclear doctrine prescribes the building of a CMD and ascribes to a no first use posture since it holds that the nuclear weapon has no role in enforcing compellence or staging aggression and, hence, is only considered usable in a situation where an adversary has first used a nuclear weapon against the country. In such an eventuality, the doctrine mandates assured retaliation to inflict unacceptable damage. In order to carry out this exercise, the doctrine aspires for a minimum nuclear deterrence whose credibility resides in its survivability.

The operational nuclear strategy as flows from India's nuclear doctrine provides the least risk option in a situation where nuclear weapons are present. It premises nuclear deterrence on a small arsenal that is not on hair trigger alert, and, hence, is less open to the possibilities of miscalculation or accidental use. At the same time, given its own orientation towards no first use and punitive retaliation in case of use, the doctrine seeks to minimise the chances of nuclear use in the first place.

Lastly, it must be reiterated that India's nuclear doctrine accords due importance to the attainment of an NWFW as the best insurance of Indian security. In an NWFW, India can be regionally more secure and globally better placed to pursue its objective of assuring strategic autonomy for its pursuit of economic and security objectives. But until such a world may be obtained, India's own brand of nuclear deterrence as defined by its nuclear doctrine will have to do the needful.

5

Hand on the Button

Nuclear Command and Control

Command and control is at the heart of credible nuclear deterrence. A doctrine conceives a role for the nuclear weapons and a philosophy for its use, but it is the command and control that must plan the build-up of the arsenal and authorise the use of the weapon, if the need ever arises. Without this system in place, the acquisition of a nuclear weapons capability would be of little significance. In fact, it is the very knowledge of the existence of a system that is *capable of weapons employment and deployment* that bestows credibility and stability to a deterrence relationship.

This chapter undertakes a comparative evaluation of the nuclear command and control structures of Pakistan, China and India. It examines the systems, as they are known to exist from the limited amount of information available on the subject in the open domain; identifies the functions of the structures in the three nations; and then proposes some specific recommendations for the Indian command and control system.[1]

Introduction

Command and control basically entails the performance of two main functions – informed decision-making, and planned execution of those decisions. The decisions to be made by a nuclear command and control system would necessarily pertain to four broad areas:

(a) development, acquisition, deployment and use of nuclear weapons;
(b) safe and reliable custody of the arsenal and its components;
(c) integration of the arsenal once the decision of use has been taken;
(d) assured delivery of the nuclear weapon on the intended target.

In order to effectively perform these functions, the basic constituents of every command and control system must include:

- A designated set of people authorised to take these decisions, whether in the primary or alternate chain of command.
- An early warning system that comprises not only technical mechanisms such as electronic intelligence, signal intelligence, satellite-based intelligence, etc but also human intelligence for holistic information gathering and interpretation.
- Command decision-making centres that have adequately built in survival measures, including through the provision of an alternate centre.

---

1. The term command and control as used in this chapter subsumes communications, computers and intelligence, surveillance and reconnaissance or the C4ISR. In the understanding of the author, command and control are the primary functions, while the other C2ISR are capabilities to assist the effective performance of these tasks. Therefore, they are components of the command and control system but not the system themselves.

- Secure and redundant communication links between decision-makers and the units that are to carry out the execution. The relay of timely instructions from the decision-making authorities to the operational commanders of nuclear forces to place them on alert, launch or recall is absolutely critical.
- Military units equipped with nuclear-armed delivery vehicles and adequately trained in assembling, maintaining, transporting, storing and operating complex systems and weapons. Besides systems and procedures, the members of these units would also need periodic personnel reliability checks.

While every command and control system would essentially have to build these capabilities, the actual operation or interaction between them would vary from country to country, depending upon the personality and culture of the military organisation, nature of civil military relations, inter-Services equations, strategic culture of the nation, alert status of the nuclear arsenal, etc.

Distinct they may be in their nature, however, command and control systems everywhere comprise a complex array of assets, personnel, procedures and information used by a decision-maker in planning, directing and controlling operations. It could well involve several hundreds of people at various levels, acting under orders emanating from different sources that are interacting with one another, horizontally or vertically, as well as with a variety of technical systems. Given the complex interconnections necessary to make the right contact, at the right time, with the right input, to reach the right decision, it is obvious that command and control poses huge challenges. In the nuclear domain, these complexities are exacerbated by the fact that the decisions, thus, made have resonance for the survival of nations. Therefore, adequate thought needs to be given to these activities, structures and functions in peace-time with the hope that everything would function smoothly during times of crisis. Of course, there can be no way of knowing how these would actually perform in case a situation was to really escalate to the nuclear level.[2] Fortunately, though, *deterrence stability* is significantly enhanced just by the *knowledge* that there is in place a command and control system that would be capable of informed decision-making even in the midst of a nuclear crisis.

The basic purpose of an effective command and control system is to exercise both positive and negative controls on the available nuclear weapons. These are enabled through a set of interconnected human, technological and administrative arrangements and procedures. Positive controls imply that the nuclear weapon is in safe and reliable custody that can *ensure its use on receiving proper authorisation.* These controls are hugely dependent on secure and redundant communication networks that can survive the strains of a crisis or actual physical damage in case of a conventional, electronic or a nuclear attack, as well as pre-designation of an alternate authorised authority to cater for decapitation. Negative controls, on the other hand, mean that weapons will *not be used unless authorised.* These may be exercised through technical solutions like the development of combination or coded locks such as the permissive action links (PALs) that can block unauthorised use of the weapon. They could also be reinforced through procedural mechanisms such as the requirement of more than one person to actually use the weapon. In the absence of PALs with the

2. Fictional writing on nuclear wars has often demonstrated the inability of C2 to perform as expected during a crisis. For instance, David Aaron's novel, *State Scarlet* (1987) brought out several dilemmas such as how to limit access to the nuclear button and yet make sure it can be pushed if something happens to the President; or is nuclear decapitation a good strategy or would it eliminate the ability to de-escalate a crisis, thereby putting "two headless chickens in a fight"?

three countries under study, each presently relies on two or three men to authorise nuclear use.

The balance between the two types of controls has to be optimum in order to avoid twin dilemmas that have long troubled nuclear planners. Peter Feaver best described these as the dilemma between making sure that the weapons do not "fail deadly" or "fail impotent."[3] He argued that if the authorities controlled the weapon too loosely, deterrence would suffer from the risk of unauthorised or accidental launch, and, therefore, could fail deadly. On the other hand, if the control was too tightly vested only in the top leadership, then a first strike had a good chance of successful decapitation, thereby reducing the probability of retaliation and making deterrence fail impotent.

Designing measures to impose positive or negative controls are not impossible tasks by themselves. The problem arises from the fact that both types of controls must be applied simultaneously and must coexist. The tension so created, forces leaders to choose either an assertive or a delegative control system. In the case of the latter, the control of the weapon and authority for its use is delegated along a structured route so that subordinate commanders can make nuclear use decisions under certain defined circumstances. This raises the possibility of unauthorised or accidental launch of the weapon in case the commander is tempted to use the weapon in situations not defined or uses it preemptively in a hasty or wrong assessment of a crisis situation. But it also provides greater assurance of retaliation and, hence, adds to the credibility of deterrence. An assertive system, on the other hand, grants tight control of the weapon only to the top political decision-makers. While this ensures negative control over the weapon, it does carry the risk of a decapitating strike leading to decision-making

3. Peter Feaver, “”Command and Control in Emerging Nuclear Nations,” *International Security*, vol. 17, Winter 1992-93, p. 160.

paralysis, unless a well-demarcated chain of succession has been prepared and is known in advance to the authorities.

Several strategic analysts have highlighted the benefits and limitations of both assertive and delegative systems, especially in the case of small nuclear arsenals.[4] A small number of nuclear weapons do warrant a more direct and assertive control to rule out accidental and miscalculated launches, but, at the same time, they are also more vulnerable to a decapitating first strike. This vulnerability then makes it necessary to have a more delegative system that can ensure retaliation and, hence, enhance credible deterrence. Similarly, while keeping the nuclear arsenal in one place reduces the burden on safety and security as also easing connectivity challenges, it nevertheless increases their susceptibility to destruction in a first strike. The dispersal of the arsenal, however, brings the opposite problem. It increases the chances of survivability of the arsenal, but also heightens the possibility of accidental or unauthorised launch, raises the requirement for reliable availability of secure communication channels, besides making them prone to theft, unless uniformly high security arrangements can be assured at every location.

Each nation arrives at its own formula to accommodate the safety and survivability concerns. These are based on their specific threat perceptions, political nature and strategic capabilities of the adversary, own technological expertise and organisational strengths. For instance, the US that during the Cold War premised its deterrence on positive controls to convey the message that its command and control system was capable of activating near-immediate nuclear use once the decision had been taken, dispersed ready nuclear weapons across diversified delivery systems to

4. For instance see Clayton P. Bowen and Daniel Wolven, "Command and Control Challenges in South Asia", *Nonproliferation Review*, Spring-Summer 1999, pp 25-35; Rajesh Rajagopalan, *Second Strike: Arguments About Nuclear War in South Asia* (London: Penguin Books, 2005).

be available with commanders for use. In order to deal with the concomitant problem of negative controls, it relied on technical solutions such as PALs and elaborate physical launch and firing sequences, as also organisational procedures such as personnel reliability programmes and the 'two man rule' – measures that were specifically designed to prevent unauthorised launches.[5] The considerations that dictate Pakistan, China and India to lean towards one or another type of command and control system are discussed in a later section of this chapter.

Pakistan's Strategic Command Organisation (SCO)

It was on February 2, 2000, that Pakistan announced the creation of its nuclear command organisation. The SCO is structurally grouped into three constituents – National Command Authority (NCA), Strategic Plans Division (SPD), Strategic Forces Commands (SFC).

The NCA is the chief decision-making body under the chairmanship of the President, with the Prime Minister as the Vice Chairman. It is responsible for policy formulation and the employment and development of strategic systems. The NCA functions through two committees – Employment Control Committee and Development Control Committee. The Employment Control Committee is chaired by the head of government and includes the ministers of Foreign Affairs, Defence, Finance and Interior, Chairman, Joint Chiefs of Staff Committee (CJCSC), the three Chiefs, Director General of the Strategic Plans Division and technical advisers. The functions of this committee include review of the latest information on the threat to the strategic weapons programme and weapons deployment. It gives policy direction

---

5. For more on this, see Ashley Tellis, *India's Emerging Nuclear Posture: Between Recessed Deterrent and Ready Arsenal* (Santa Monica: RAND, 2001), pp. 429-430.

and guidance for the evolution of doctrines and employment policy based on technical capability and threat assessment. It also establishes the hierarchy of command, formulates policy for delegation of authority for employment of nuclear weapons; and establishes guidelines for an effective command and control system to safeguard against accidental or unauthorised use.

On the other hand, the Development Control Committee is also chaired by the head of the government and comprises the same military and technical members as the Employment Control Committee minus the Cabinet ministers. However, it includes the representatives of the weapons research, development and production organisations such as the A.Q. Khan Research Laboratory, National Development Complex and Pakistan Atomic Energy Commission. It exercises technical, financial and essential administrative control over the strategic organisations and oversees the systematic development of strategic weapons programmes as per approved development strategy.

The SPD consists of the Operations and Planning Directorate, a Strategic Weapons Development Directorate, Arms Control and Disarmament Affairs Directorate, Computerised Command, Control, Communications, Information, Intelligence and Surveillance Directorate, Consultancy Directorate, Security Division and Strategic Forces Communications Planning Cell. These provide secretarial support to the NCA and handle all issues related to the nation's nuclear capability. It functions directly under the President, Prime Minister and Chairman, Joint Chiefs of Staff Committee. Headed by a general from the army, it comprises officers from the three Services. Its main functions include:

- Formulation of the country's nuclear policy, strategy and doctrine, as well as arms control and non-proliferation policies.
- Formulation of short-and long-term development strategy and force goals for tri-Services strategic forces.

- Liaising with all strategic organisations as well as supervision of their budgetary and administrative aspects.
- Daily management of strategic assets, including their safety and security .
- Establishment of lower tiers of the command and control system and their physical infrastructure.

Separate SFCs are in place in all the three Services. The Services retain training, technical and administrative control over their strategic forces. The strategic operational policy framework and guidelines for employment of nuclear weapon systems are established by the NCA through the SPD. Based on these guidelines, the three Services formulate their operational policies, which are meshed and harmonised at the tri-Service level by the SPD in coordination with Director General Plans of the Joint Service Headquarters.

Pakistan's command and control system may be described as being less assertive than those of India and China since it envisages delegation of control for nuclear use far more than the Indian and Chinese systems. This is not entirely illogical given that Pakistan is conscious of its lack of strategic depth that makes its nuclear assets and command structure more vulnerable to an Indian strike. Though India has declared an NFU policy, Pakistan has been sceptical of the declaration and seeks to strengthen its deterrence with the help of a more delegative system.

The army plays a very important role in the Pakistani command and control. It has always been the primary decision-maker on matters nuclear and this reflects in the command and control structures too. This factor also makes it easier for Pakistan to adopt a delegative posture since pre-delegation of authority to field commanders is consistent with the overall military style of functioning. Pakistan is not yet credited with having a sophisticated early warning system.

Fig 5.1

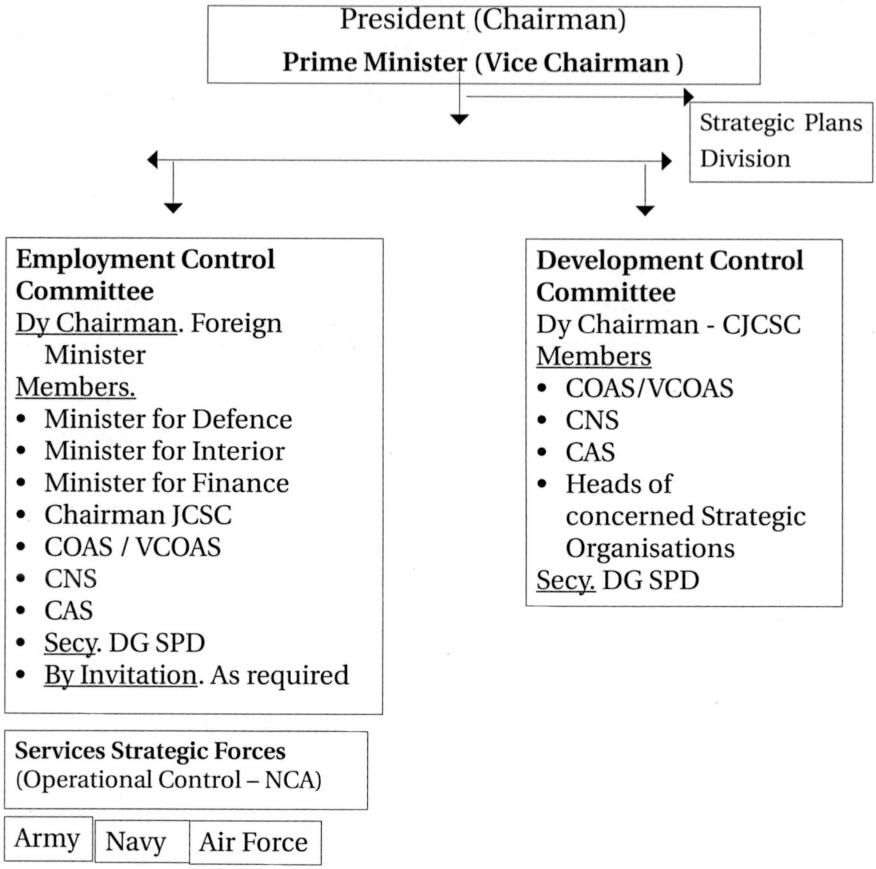

China's Second Artillery Corps

In a speech on the occasion of the 40th anniversary of the founding of the Second Artillery Corps (SAC) in 2007, President Hu Jintao described it as a "strategic force directly commanded and used by the Party Central Committee and the Central Military Commission and is our core force for strategic deterrence."[6] Significantly, the

6. Larry Wortzel, *China's Nuclear Forces: Operations,Training,Doctrine,Command, Control and Campaign Planning* (Carlisle Barracks, Pa: US Army War College, Strategic Studies Institute, May 2007).

SAC, also known as the Strategic Rocket Forces, is not a part of the PLA Army, Navy or Air Force.

The Commander-in-Chief of the SAC is the top most echelon at the operational level and is networked with the command organisations of the military departments, missile bases and the firing units. He is responsible for the plans to execute the dispersion of the missile units in times of crisis and controls the firing orders. The commander, however, reports to the General Staff Department (GSD) of the PLA which in turn gets its orders from the Central Military Commission (CMC), the highest and the most centralised level of military leadership in the Chinese Communist Party. The SAC commander is also a member of the CMC, thus, providing a larger role for the SAC in the military and strategic planning process.

The SAC consists of an early warning division, a communication regiment, a security regiment, a technical support regiment and six ballistic missile divisions. The SAC is headed by a general, and the six bases by a major general each. There are15 missile brigades under the six bases and each brigade that operates one type of missile is headed by a colonel. Highly centralised, redundant and networked command and control system exists for the Chinese nuclear forces. SAC orders are believed to be encoded and protected, requiring human authentication. There is an emphasis on "man in the loop" for nuclear firing orders and, in that sense, the command and control does not exist as a fully automated system. However, with the current emphasis on informationalisation and the growing familiarity with computers, would this approach sustain? Today, the SAC also controls conventional missile forces alongside the nuclear missiles.

Fig 5.2

China's Operational Nuclear Command and Control Organisation

Chinese Communist Party Politburo
Chairman – General Secy

↓

Central Military Commission

↓

General Staff Department
Headed by a General also
member of CMC

↓

Second Artillery Command HQ
Also member of CMC since 2004

↓

6 Bases/Divisions
Each headed by a Major General

↓

15 Brigades
Each headed by a Colonel

India's Nuclear Command and Control

Though the nuclear doctrine crafted by the first National Security Advisory Board laid out the need for a robust command and control structure for the establishment of Indian nuclear deterrent, the actual process of its organisation has been slow and largely hidden from public purview. There has been a general hesitation to accord too much importance to this aspect possibly because of the widespread belief that nuclear weapons are essentially political instruments of deterrence, warranting no real need for a command and control system to actually make operational plans for nuclear war-fighting. This assumption, however, ignores the inherent paradoxical logic of nuclear deterrence that demands the projection of readiness to use the weapon in order to make it credible and sustainable. Therefore, a more focussed approach to the subject is necessary.

Within months of the nuclear tests in May 1998, a National Security Council (NSC) was established in November of that year. Headed by the Prime Minister, it included Ministers of Defence, Home, External Affairs and Finance. The principal secretary to the Prime Minister was designated the National Security Advisor. A Strategic Policy Group was established to complement the functioning of the NSC and it comprised military and civilian officers. These included the Cabinet Secretary, the three Service Chiefs, Secretaries for Home, External Affairs, Finance Ministries, Secretary for Defence Production and Atomic Energy, Space and Revenue. This body was tasked with the conduct of a Strategic Defence Review drawing upon inputs received from the National Security Advisory Board (NSAB) consisting of former civilian and military officials and analysts from the fields of science and technology, economics and strategy.

Within a few years of the tests, a spate of writings by strategic analysts studied and considered alternative models for the command and control structures. A lot of attention was devoted to the need for creation of the post of Chief of Defence Staff (CDS) who would actually be responsible for the command of nuclear weapons and provide a direct single channel of advice and communication to the Prime Minister or the Political Council. However, the idea was, and remains, engulfed in controversy given the disagreements within the Services on who should be the CDS and whether there was requirement of such a post at all since modern complex warfare scenarios require multiple inputs on matters as critical as nuclear.[7] In any case, the idea placed before the government as part of the report of Group of Ministers stands rejected by the Prime Minister and the Cabinet Committee on Security (CCS) for the time being.

Subsequently, in January 2003, in a note seeking to "operationalise" the nuclear doctrine, the CCS mentioned the creation of the National Command Authority (NCA) comprising a Political Council (PC) and an Executive Council (EC). The former is chaired by the Prime Minister and is the sole body to authorise nuclear use. Meanwhile the Executive Council is chaired by the NSA and provides inputs into decision-making of the Political Council and also executes directives received therefrom. The CCS also approved the appointment of a Commander-in-Chief, Strategic Forces Command (C-in-C, SFC), to manage and administer command and control over nuclear forces. He is responsible for training and other operational issues related to nuclear weapons. Of course, in the Indian system as it prevails, the responsibility for the safe storage and reliability of the fissile core rests with the Department of Atomic Energy (DAE), of the weapon assembly with the Defence Research and Development Organisation (DRDO), and the delivery vehicles are maintained and controlled by the individual armed forces. A National Command Post, as also an Alternate Command Post is catered for.

---

7. For a viewpoint supporting the creation of the CDS, see Gurmeet Kanwal, *Nuclear Defence* (New Delhi: Knowledge World, 2000), pp. 159-161. For a contrarian view that argues in favour of the Chiefs of Staff Committee to be the nodal organisation for defining nuclear requirements, making targeting plans and transmitting nuclear release orders, see Raja Menon, *A Nuclear Strategy for India* (New Delhi: Sage Publications, 2000), pp.248-251. Air Cmde Jasjit Singh (Retd), meanwhile, refutes the need for 'single point advice' to the government and compares the conduct of modern warfare with the conduct of an open-heart surgery where different medical specialists must work as a team to save the patient.

Recommended Structure for India

As is evident from what is known about the existing structure of the Indian nuclear command and control, there is room for improvement, especially in terms of transparency and organisational clarity, including in composition and functions. It is quite likely that the manner in which the nuclear command and control system has been allowed to develop is a deliberate decision, indicating a lack of desire to tamper with the existing informal arrangements that have existed. Quite different from the Indian command and control system, its counterpart in Pakistan, at least projects greater lucidity and sense of purpose. Meanwhile, from the little that is known of the Chinese nuclear command organisation, it stands out for its linearity and clarity of echelons, and, hence, appears better geared for action in times of crisis. This perception is further enhanced by the better integration of civilian and military leadership in the command structures of Pakistan and China. But then all of this is a reflection of the larger political reality and the nature of political systems of the three nations. Two major aspects of the Indian command and control system particularly stand out as compared to others.

One, the nascent Indian nuclear command and control system has traditionally existed as a more informal and loosely defined set of arrangements. Leaders of the concerned organisations have found it easier to reach across horizontally and vertically to perform their functions. This, in fact, has been especially true of cross-organisational contacts such as, for instance, between the military and the DAE or DRDO, etc. Within each organisation, though, structures and formalised procedures for functioning have existed. This is described by Ashley Tellis as "a good example of how nested 'organic' arrangements can always be embedded within, and derive their viability from otherwise large and complex formal organizations."[8] Of course, informality of arrangements also enabled greater secrecy, which was necessary for tasks that needed to be done with minimal visibility, employment of resources, or alteration of pre-existing patterns of activity in the pre-1998 period.

However, it must be highlighted that informal systems, for all their advantages, will not be adequate to handle the complexities that will increase as India's nuclear capabilities and arsenal grows, delivery systems diversify, and dispersal of nuclear assets demands greater coordination and planning. The need to rapidly alert, relocate, integrate and ensure readiness of a larger and more diverse nuclear force would necessitate a more institutionalised and structured command and control to grapple with challenges of targeting analysis, contingency planning, integration of components available with different custodians and synergisation of ancillary conventional capabilities in support of nuclear operations.

Secondly, in the Indian system, as per the present composition of the NCA, the armed forces remain out of the nuclear decision-making loop, unless specifically invited for advice or consultations. This cannot be an optimum model for credible deterrence. In any case, no decision on nuclear use is possible without receiving adequate inputs from the actual practitioners of this capability and, hence, a more formal integration of the military, as recommended in the accompanying model, would be appropriate. While it is necessary that the ultimate authority for nuclear use must vest in the political head of the government, namely the Prime Minister in the case of India, or his successor(s) as designated by him through a pre-determined chain of succession, the institutionalised presence of the military leadership in the PC, and their regular interaction in periodic meetings/briefings would enable informed and quick

8. Tellis, n. 5, pp. 451-452.

nuclear decision making at the time of crisis.

Before making specific recommendations, it would be necessary to evaluate the threat scenario for India and its implications for the nature of command and control. How assertive or delegative must it be? What kind of trade-offs can the system accommodate between safety and survivability of the arsenal? What is the level of threat of nuclear use, both deliberate and inadvertent from the adversaries? For an Indian command and control system to be responsive, survivable, comprehensive in constitution, and better integrated at all levels, it is imperative that adequate thought be given to these questions.

First, unlike the situation of the superpowers during the Cold War which were always fearful of a "bolt out of the blue" kind of nuclear attack and, hence, placed greater emphasis on immediate response, preferably on the first detection of launch of enemy missiles, the Indian system can afford to be more relaxed on that front. In the case of India, the remote possibility of nuclear use by Islamabad or Beijing would happen only after a crisis situation has developed over a period of time. In any case, India and China have a declared no first use policy and, hence, do not envisage a direct escalation to the nuclear level. With Pakistan and its first use posture, it tries to project a low nuclear threshold. However, given that Pakistan's first use would invite certain punitive retaliation, Islamabad would also be wary to use the weapon except as a measure of last resort. Therefore, for India, the likely escalation to the nuclear level will happen once it is well into the conventional phase of operations. This has two implications: one, the level of alert of the nuclear arsenal will be determined by the unfolding conventional crisis; second, the dispersed elements of the nuclear arsenal will have to be congregated even as the conventional operations are on. This could be a cause of concern for security of assets and their being attacked inadvertently while in transit. It is critical that these scenarios not only be anticipated and planned for by the military, but that the political leadership is also familiarised with contingencies in which decisions of the nuclear kind are quickly required. Of course, the Indian command and control can afford to be less time sensitive compared to the predicament of the superpowers. Nevertheless, retaliation must take place within hours, as few as possible, if the country is to avoid action paralysis owing to international pressure, or being subjected to a possible second wave of nuclear strike.

An NFU policy, premised on a retaliatory punitive strike, does not require an operationally deployed arsenal to be kept at hair trigger alert. This, as mentioned earlier, makes it possible for India to maintain its nuclear arsenal in a disaggregated form at different locations. Therefore, the DAE has the custody of the fissile core, the DRDO of the non-nuclear weapon assembly, and the military of the delivery vehicles. While this organisational and, hence, physical separation of components contributes to negative controls, it does, however, complicate the command and control system. For one, it requires a larger number of people in the nuclear loop at the operational level. Secondly, it has implications for the communication networks that must be available to every one of the actors because in the absence of connectivity between even one of the segments, deterrence would carry the risk of failing impotent.

While disassembled in form and with multiple agencies exercising custodial control, the Indian nuclear arsenal, however, could be strongly centralised in command. It has been argued that since the Indian nuclear weapon is envisaged to play only a political role of deterrence, there is no need for pre-delegation of authority for nuclear use. Rather, this would be vested in the person of the Prime Minister and transferable only in conditions of emergency. In order to ensure credibility of deterrence by evoking certainty of retaliation in the presence of such an assertive model of command, it becomes imperative to take measures that can reduce the vulnerability of strategic decapitation. The system, therefore, must be built to accept and sustain loss of leadership in order to "ensure a continuing capability to effectively employ nuclear weapons" even in the face of nuclear attacks and/or repetitive attrition attempts.[9] This obviously calls for the development of a proper, pre-determined system for devolution of authority in an emergency. It should allow a seamless and almost automatic transfer of authority to survivors in case of an attack to order nuclear use in accordance with established and publicly recognised procedures. Not much is known on whether the Indian command system has a pre-designated chain of succession. By all indications, it should. In any case, the January 4, 2003, statement of the Cabinet Committee on Security (CCS) on "Operationalisation of India's Nuclear Doctrine" had provided some pointers on this in its rather cryptic statement,"The CCS also reviewed and approved the arrangements for alternate chains of command for retaliatory nuclear strikes in all eventualities."[10] Greater transparency on the fact that adequate procedures have been established in this regard without necessarily identifying positions in the chain would go a long way in enhancing the credibility of deterrence.

Based on the need for satisfying some of the conditions highlighted in the paragraphs above, a recommended nuclear command and control structure for India is graphically depicted in the Fig 5.3 below. It retains several features of the existing organisation. But it also proposes slightly altered membership or the creation of new bodies to perform certain specific tasks.

Recommended Nuclear Command and Control for India

9. Draft Report of the NSAB on Indian Nuclear Doctrine as presented to Government of India, August 17, 1999. Reproduced in Appendix 1 of the book.
10. Note by CCS on Operationalisation of India's Nuclear Doctrine, January 4, 2003. Reproduced in Appendix 2 of the book.

Fig 5.3

Recommended Nuclear Command and Control for India

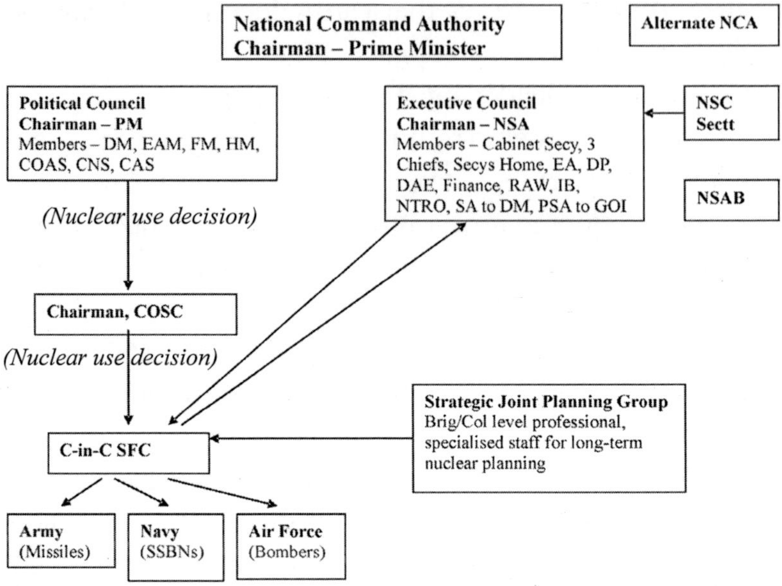

Legend:

DM	– Defence Minister
EAM	– External Affairs Minister
FM	– Finance Minister
HM	– Home Minister
COSC	– Chiefs of Staff Committee
NSA	– National Security Adviser
COAS	– Chief of Army Staff
CNS	– Chief of Naval Staff
CAS	– Chief of Air Staff
DAE	– Department of Atomic Energy
RAW	– Research and Analysis Wing
IB	– Intelligence Bureau
NTRO	– National Technical Research Organisation
SA to DM	– Scientific Adviser to Defence Minister
PSA to GOI	– Principal Scientific Advisor to Government of India
NSCS	– National Security Council Secretariat
NSAB	– National Security Advisory Board
C-in-C SFC	– Commander in Chief, Strategic Forces Command

In the structure recommended above, the National Command Authority with the Prime Minister as the chairman is the apex body for managing the nuclear capability of the country. It shall comprise two constituent organisations: the Political Council (PC) consisting of the Prime Minister and the Ministers for Defence, External Affairs, Finance and Home, *besides the three Chiefs of the armed forces*. They would bring the military input into the political decision-making apparatus of the NCA, which is currently missing. Traditionally, the political leadership has been hesitant to bring in military members into this council for fear of according a greater role to them. Militarisation of nuclear policy has been consciously avoided in order to preclude the possibility of the military demanding excessive resources, including perhaps, the transfer of the physical custody of the weapon. However, it must be conceded that the military is well aware of the complete political control over the nuclear arsenal, as also other military affairs, in a parliamentary democracy. The Services realise they would not receive the custody of the nuclear weapon until such time as the political leadership has decided, as a crisis unfolds. But, decision-making on nuclear use, which is the primary function of the PC, can hardly be performed without keeping the three Chiefs in the loop. Given the presence of nuclear weapons with India and its adversaries, it is inevitable that their shadow would fall on every conventional — limited or mountain war — that might break out in the region. The close inter-linkage between conventional war and nuclear deterrence, as explained in Chapter 10, makes it imperative that the interface between the political and military leadership must be far more than the present Political Council envisages. In fact, the presence of the military members in the PC would serve two purposes: one, it would enhance the perception of credibility of deterrence by the adversary; two, it would increase the understanding of both the leaderships — civil and military— of each other's compulsions and preoccupations. With modern wars certain to be politico-diplomatic-military exercises more than ever before, this comprehension would serve a larger purpose of synergy of thought, planning and effort.

The PC must meet regularly[11] to receive briefings from the NSA or other invited experts on various aspects of security contingencies. These would go a long way in sensitising key Cabinet members on a range of security issues, encouraging greater understanding and reducing response time in the face of a crisis. Given the nature of threats before India, especially the growing probability of nuclear use by a non-state actor, surreptitiously supported by the Pakistani establishment (or some of its elements), response options need to have been thought through for rapid reaction to the situation. The formulation of several contingency plans (for state and non-state nuclear use) and their familiarisation to all members of the PC, and making this fact public, would help in reassuring own populace while simultaneously sending a signal to the adversary and thereby increasing deterrence credibility. This would also help anticipate a situation where, in an emergency, the PC is unable to convene (because of the physical unavailability of some ministers). The Prime Minister, called upon for a quick decision, can then draw upon these peace-time discussions and deliberations in which his colleagues would have already shared their thoughts on similar contingencies.

The second arm of the NCA would be the Executive Council that would be chaired by the National Security Advisor. Its members would include the Cabinet Secretary, the three Chiefs as well as the Secretaries of Home, External Affairs, Defence Production, DAE, Space, Revenue, RAW, IB, Scientific Adviser to the Defence Minister, and the Principal Scientific Adviser to the government.

11. The Kargil Review Committee headed by K. Subrahmanyam had advised the CCS (which effectively is the existent PC) to meet every fortnight.

The main functions of this body would include undertaking scenario building on the basis of considered threat and capability assessments of the adversary and sharing these regularly with the PC. The National Security Council Secretariat would provide secretarial support to the Executive Council, including periodic assessments of threats, nuclear trends of the adversaries, their nuclear posture and other evaluations such as their threshold of unacceptable damage. These must be undertaken on the basis of deep and extensive study of the cultural, socio-political and strategic factors affecting the adversary. Meanwhile, the NSCS would itself draw upon inputs received from the National Security Advisory Board (NSAB) comprising retired bureaucrats and military officers, academics and corporate leaders, as well as the expertise of their own professionals.

The operational control of the nuclear forces would be vested in the Commander-in-Chief, Strategic Forces Command who would receive the order authorising nuclear use from the Political Council in times of crisis through the Chairman, Chiefs of Staff Committee (COSC). Unlike the Strategic Air Command of the USA or the Second Artillery Corps of China, the tri-Service SFC, would not centralise custodial and maintenance responsibilities of nuclear weapons. In fact, custody of the weapon would remain distributed between the DAE and the DRDO as mentioned earlier. Maintenance of the weapon would also be their responsibility while the same of the delivery vehicle would be undertaken by the respective Services. The C-in-C, SFC would exercise control over the strategic units either through direct command or through the three Services which would actually store, maintain and deploy the delivery mechanisms. It is recommended that the dual role delivery systems be physically maintained by individual Services, with those allocated a purely strategic role to be available for command of the SFC. As road and rail mobile missiles get inducted into the forces, and SSBNs become operational, the SFC would be responsible for their dispersal, concealment, random movement and relocation mixed adequately with dummies in order to enhance the chances of survivability.

Therefore, the main function of the SFC would be more in terms of planning and coordination to conduct training and exercises in peacetime and arsenal integration in crisis situations. This would include functions such as prior route planning for aircraft and pre-planned allocations and organisational arrangements of supporting systems such as escorts, tankers, jammers and airborne early warning command and control platforms, etc. As is evident, these operations would require a close interaction among the necessary components of the three Services and the need for joint planning and exercises cannot but be overemphasised.

There is also a case for creation of a permanent strategic staff to perform long-term perspective nuclear planning. This body that may be called the Strategic Joint Planning Group (SJPG), which would be a tri-Service body to cater for the planning, development and deployment of nuclear assets in a joint manner. It would comprise military and scientific professionals and could be an adjunct of the SFC. Therefore, the SFC would have an operational and a planning division. The SJPG would be populated by officers of the rank of colonels and brigadiers and their equivalents in other services, DAE and DRDO, as well as other working level staff. It would be treated as a specialist organisation to undertake policy planning on strategic force structure, developing targeting plans, planning dispersal of assets, and contemplating other survivability issues. Such planning requires greater collaboration between the military, DAE and DRDO to evolve coordination procedures for the custodians and the users. The SJPG would enable this as the nodal agency for planning and evolution of policies/procedures. The deliberations and recommendations of the SJPG may be taken by C-in-C, SFC to the EC and through the NSA to the PC for approval. Such an institutionalised professional body tasked with future planning would facilitate evolution of well-considered policies. Joint planning for new nuclear acquisitions/inductions and the allocation of existing nuclear designated assets even when the delivery vehicles are stored and maintained by individual Services would help in avoiding duplication and fostering better coordination and more effective operations.

While all necessary measures need to be taken to increase the survivability of the National Command Authority, it is equally critical that an Alternate NCA is created to replicate effective discharge of the same responsibilities of the NCA in the case of a decapitating first strike. Given India's no first use posture, this is of utmost importance because if the command and control does not survive a first strike, then the survival of other nuclear assets would be meaningless. Adequate attention must, therefore, be devoted to building a clear chain of succession after the Prime Minister on the civilian side and the Chairman, COSC on the military side. While the order of succession may be classified, it is imperative that the adversary be aware that such provisions have been made. Even more importantly, the designated ministers should be periodically educated on the threat assessments and unique aspects of nuclear deterrence for them to be able to effectively perform in a case of crisis.

Future Challenges

Command and control structures and procedures have to be dynamic and responsive to the changing threat assessments and own capabilities. Therefore, these cannot be cast in stone. As the nuclear force structure changes as a result advancing technological capabilities, so will the nature of their management. Among the challenges that are likely to emerge before the Indian command and control, the following deserve to be mentioned.

Emergence of New Class of Threats

Until now, the nuclear threat has largely been envisaged in the context of deterrence breakdown between nuclear-armed states. C2 has, therefore, been tailored to anticipate and respond to such exigencies. However, for India, a potent threat could also emerge from the use of the nuclear weapon, or an RDD, by a non-state actor serving as a proxy for a state. It is not unimaginable that some radical elements in the Pakistani government may encourage and support a terrorist outfit to smuggle a nuclear weapon or material into India through the sea route (in a container), over land borders from any of India's neighbouring states, or even bring it in by civilian aircraft originating from Pakistan or one of its friendly nations. A nuclear detonation on Indian soil, without the use of a missile, could then even be alleged to have been an accident involving an Indian nuclear weapon or the handiwork of some Indian groups of disaffected people. Pakistan, in other words, would be able to maintain its façade of "plausible deniability". It may be mentioned that such a scenario has been generated in *Line of Control*, a recent work of fiction, in which Pakistan uses terrorists, without even their knowledge, to bring in a weapon over the mountains, with the help of mules.[12]

It may not be prudent to dismiss this scenario out of hand. Rather, it calls for a careful consideration and assessment in order to anticipate the unthinkable. The EC and the PC must be apprised of such possibilities and responses must be thought through. Should India treat this as first use and accordingly prepare for nuclear retaliation, or can other responses suffice? This knotty issue deserves a debate. Meanwhile, India must plan for round

12. Mainak Dhar, *Line of Control* (New Delhi: Vitasha, 2008).

the clock surveillance of Pakistan, in time through a number of satellites. Effective intelligence, border and port controls, and speedy investigation would have to be the other pillars to buttress prevention of such acts.

Induction of New Capabilities

The transformation of delivery systems with the induction of nuclear submarines as compared to a force structure that earlier primarily relied on air delivery would obviously require adjustments in the command and control systems. While land-based nuclear delivery capability allows the arsenal to be kept in a disaggregated form, SSBNs presuppose a ready, complete nuclear weapon system being available with the commander of the vessel. This obviously has implications for the predominantly political control of the nuclear weapon and introduces greater strains of delegation in an otherwise assertive command and control system.

Also, increasing mobility of the nuclear arsenal, with the induction and deployment of road and rail mobile missiles, would require the command and control systems to coordinate frequent moves and relocation of assets. This would lead to the urgent need for developing technical capabilities, human competence, infrastructure support and organisational arrangements to integrate weapons systems as well as adequate safety and security measures at all sites. This challenge will remain in the future because, with the exception of the few missile systems dedicated solely to nuclear use, other war-fighting systems such as combat and transport aircraft, communication equipment, surveillance and battle damage assessment assets will be dual capable.

Need for Reliable, Redundant and Secure Communications

Reliable, redundant and secure communications would always

remain the Achilles heel of command and control systems. These are the weakest links and suffer from the highest levels of vulnerability given the use of weapons such as EMP bombs close to ground stations or possible future threats to communication assets in space. Therefore, the need for multiplicity of communication systems would always remain. Of course, space-based communications have advantages that must be exploited including those of wide bandwidth, high data rate communications between multiple users in different media. These would also enable audio-video conferencing for the conduct of military operations. Today, dedicated military communications satellites are within India's technical capability, but the vulnerabilities of the same to jamming, destruction of ground infrastructure, etc must be considered. As long as these can be hardened, there would be little by way of physical threat from an anti-satellite weapon to these assets since they are stationed in the geosynchronous orbit, hence, out of range of being harmed physically by ASAT weapons.

Devotion of Adequate Time by Political Leadership

It has been reiterated through the book that nuclear weapons are political tools and that in the case of India, this instrument is completely in the hands of the political leadership. Therefore, there is dire need to increase understanding of nuclear decision-making by the political leadership in order to enhance the credibility of deterrence. This will remain a critical challenge of command and control because only when the political leadership is able to find the necessary time to comprehend the complex requirements of credible deterrence – in show as well as action – can it become an effective player in the game of deterrence. The problem, however, lies in the fact that the Indian political system demands a lot of time of the leader for public dealings and party activities. In fact, since coalition politics is likely to be the norm in the coming years, political adjustments between parties will take up even more time of the political leadership, thereby leaving the leaders with less time or inclination to attend meetings of the PC or other briefings on nuclear matters. However, the political controllers of the nuclear weapon must be made to understand the prerequisites of credible deterrence and how the game of signalling can be effectively used to enhance deterrence.

Conclusion

The mere demonstration of a nuclear weapons capability means little for credible deterrence. The nuclear weapon must be accompanied by a range of paraphernalia for it to credibly prevent nuclear war. Command and control is one such prerequisite. It is the body that clarifies the role of nuclear weapons in national security, determines the adequate nuclear force structure, and ensures its use when authorised and its safe custody otherwise. Each NWS arrives at its own mix of positive and negative controls to build a C2 system best suited to its requirements in terms of external threat perceptions and internal power dynamics. Personalities of decision-makers have traditionally played a large role in moulding the character of the organisation of command and control systems.

The Indian nuclear command and control system has largely evolved out of the existing loose set of inter-relations and personal arrangements that had been informally worked out during the long period when India maintained an open nuclear option. Skewed heavily in favour of an assertive political control, there was little participation of the military in the nuclear decision-making during this period. This arrangement was adequate and acceptable at a time when India had not declared itself a state with nuclear weapons. However, once having done so, the establishment of the credibility of nuclear deterrence demands a far greater role for the military. Even though there is a consensual Indian view that the nuclear weapon is a political instrument to exercise deterrence and not for fighting a war with, the military is yet required to portray certain preparedness in order to ensure that deterrence remains stable and sustainable. A well structured command and control organisation that projects clear and robust inter-linkages between the various players can go a long way in conveying this message.

6

Tricks of the Trade

Survivability of the Arsenal

Existential nuclear deterrence is derived from the basic reality of the existence of nuclear weapons – irrespective of their yield or numbers[1]. The mere fact that there is some sort of nuclear capability that can impose a punishment that may be higher than the value of the benefit sought is believed to be enough to deter. Deterrence practised by North Korea against the USA is an example of this. Through the conduct of a nuclear test in October 2006, however imperfect or unreliable, Pyongyang has

---

1. There are several interpretations of the term "existential deterrence." Its first use is attributed to McGeorge Bundy who opined, "as long as each side has thermonuclear weapons that could be used against the opponent, even after the strongest possible preemptive attack, existential deterrence is strong..." But there are other versions of the term, such as by Marc Trachtenberg that premise it on "the mere existence of nuclear forces," which is enough to create a fear of escalation that must always be factored into political calculations. Still others like Devin T. Hagerty suggest that existential deterrence works even in the absence of openly acknowledged nuclear forces, as long as the adversaries believe that the opponent has nuclear forces. The term as used in this book leans closer to Trachtenberg's definition of existential deterrence. Bundy's definition of the term, in the view of the author, corresponds better to credible nuclear deterrence as used in this chapter.

managed to inject a seed of doubt and uncertainty in the mind of Washington, thereby raising risk, complicating US calculations, and constraining its actions. Existential deterrence suffices in the case of the Democratic People's Republic of Korea (DPRK) for two reasons: one, because it seeks to deter the United States, a country that is perceived to have a low damage tolerance threshold; and second, because from the American perspective, the stakes in any conflict with DPRK would never be high enough to justify any loss, however limited, that America could suffer from DPRK's nuclear use. Therefore, for Pyongyang, the power of mere suggestion of presence of nuclear weapons is good enough to impose deterrence.

None of the other nuclear deterrence relationships, however, has been satisfied with existential deterrence, except in the very early years after the acquisition of their nuclear capability. Thereafter, each one of them has built, or at least aspired to make deterrence more *credible and stable* by developing capabilities, establishing systems, adopting procedures, and evolving organisations that are capable of mounting *convincing* threats of *assured* nuclear use, if it ever became necessary. The attempt has been to ensure that chances of deterrence breakdown, or the possibilities of use of nuclear weapons are minimised, if not completely obviated, by communicating to the adversary that there are capabilities and strategies in place that would not only prevent him from achieving his objective, but also cost him dear.

The imposition and sustenance of this type of nuclear deterrence, one that is credible, hence, more likely to be stable, in contrast to merely existential, is particularly important for a country like India that has territorial conflicts with its eastern and western neighbours, both of which are nuclear armed. Given the almost pedestrian possibility of 'routine' border skirmishes escalating to a nuclear exchange as a result of deliberate choice, accident, or miscalculation, reliance on the mere psychological assurance of existential deterrence cannot be enough to prevent deterrence breakdown. Therefore, enhancing the credibility of deterrence in order to make it more stable, and, hence, less prone to meltdown, is the basic endeavour of India's nuclear strategy. This is sought through the extrapolation of a *lucid nuclear doctrine* that clearly identifies a narrow, political role for India's nuclear weapons; through the establishment of a *precise command and control* system that minimises the risk of nuclear exchange due to miscalculation or unauthorised use; and by ensuring *assured retaliation* in case of a nuclear attack in order to compel the adversary to well calculate the dangers of his nuclear use.

The strategy of assured retaliation seeks to impose deterrence by punishment. The credibility of deterrence in this case is predicated on the communication of the absolute certainty that nuclear use by the adversary would be met with retaliation that would cause damage of the kind the adversary would find unacceptable. For this to happen, an essential prerequisite is the existence of sufficient amount of nuclear capability. In the case of a first use doctrine, a nation would have the option to choose what to target and how much. But even then, for deterrence to function even after the first use, it becomes contingent upon the country undertaking the first strike, which it hopes would be sufficiently degrading if not completely disarming, to ensure that it can make survivable a sufficient capability even after absorbing a counter-strike. This implies making the nuclear arsenal capable of avoiding, repelling, or withstanding an attack in order to be available for a second strike. The first strike, however expansive or weak, can hope to deter a counter-strike only if it is adequately supported by the ability to conduct another wave of retaliatory strikes that would wreak even further damage on the adversary. The US realised this as Soviet nuclear forces grew, adding to their worries about the vulnerability of their nuclear missiles and bomber bases to a Soviet first strike. This threatened to degrade their ability to mount "massive retaliation" to cause assured destruction because of the possibility that not much might survive a Soviet first strike. This affected the credibility of the threat of punishment because if the forces that were to carry out the act of punishment were not going to survive, then the threat lost credibility.

In order to buttress deterrence in this situation, emphasis came to be placed on building a second strike capability or the capacity to survive an attack sufficiently to deliver devastating retaliation. Three approaches were adopted to achieve this objective. One, the US went into an overdrive of vertical proliferation, believing that the more the numbers in the nuclear arsenal, the greater the chances of their survival. The folly of this approach was either not evident, or ignored, in the arms race that ensued between the superpowers. Second, greater priority came to be accorded to the technical systems necessary for launch-on-warning (LOW) postures so that in case of a detection of a nuclear attack, the response would be nearly automatic ensuring, thereby, that the arsenal would not be destroyed before it launches itself. The third approach to making the nuclear weapons survivable was to harden their storage sites and make the launch platforms mobile. Hence, the shift to silos and mobile delivery vehicles, including submarines capable of launching nuclear missiles.

As is evident from the above experience, a country with a first use nuclear doctrine found its deterrence to be credible only when supported by a second strike capability (drawing lessons from this nuclear history, Pakistan is engaged in building such a capability) that could assure the nation that a sufficient amount of nuclear capability would survive even after the first nuclear attack. The logic behind this was to not only use his first strike to eliminate or degrade or dilute a retaliatory counter-strike, but also further deter the adversary from using what he had left by suggesting that the first user would still have enough to cause further damage even after the adversary's retaliation.

If credibility of deterrence in the case of a first use doctrine relies on the ability to ensure survivability of sufficient retaliatory capability to undertake a second strike, this is even more critical in the case of a no first use or counter-strike doctrine. In fact, once a country has committed itself to NFU, attention and energies need to automatically shift to making the nuclear arsenal survivable because in this case, the credibility of deterrence is entirely determined by the ability of surviving nuclear forces to accord punishment. What exactly does this entail? Which are the specific elements of the nuclear arsenal that need to be made survivable? How best can survivability be ensured? How much of the arsenal must be made survivable? This chapter takes an indepth look at these survivability challenges in the context of India's nuclear strategy.

What Should be Made Survivable?

The success of a counter-strike nuclear strategy is based on the communication of a clear message that any use of nuclear weapons would trigger assured punitive retaliation to cause "unacceptable damage" upon the attacker. If this message has been conveyed and understood properly, then it should be assumed that the aggressor would want to strike and neutralise those Indian capabilities that would enable India to mount a counter-strike. In other words, it would seek to degrade India through a disarming strike so that India is unable to retaliate. In order to accomplish this, the adversary would attempt to hit at one or both of two kinds of targets. One of these would be the country's nuclear forces such as weapon storage sites, missile silos, submarine and bomber bases, command and control nodes, etc., in order to degrade the retaliatory capability. The other target would be the nation's political will to retaliate. The aggressor would seek to disarm the country of its will by undertaking counter-value strikes in the hope that the politico-psychological impact of nuclear attacks on population centres could paralyse the leadership into inaction, thereby reducing the chances of retaliation.

However, if the credibility of deterrence is to rest on the certainty of retribution, then India needs to make these very elements survivable to ensure that it is not possible for the aggressor to degrade its capability to retaliate even after a devastating first strike. Rather, the signals to the adversary must convey that the chances of his being able to carry out either a disarming or a decapitating strike against India are close to zero, thereby disabusing him of any notion of a "splendid first strike." The message that must be communicated is that however splendid the strike might be in keeping with the adversary's capabilities, it would nevertheless not be able to guarantee destruction of India's retaliatory wherewithal. Such a perception of survivability would significantly enhance the credibility of deterrence not only by reining in the adversary's temptation for a first strike, but also by tilting the balance in favour of non-use of nuclear weapons.

For the above to translate into reality, however, measures towards increasing the survivability of the nuclear arsenal must be pursued in a systematic and planned manner. This first calls for an identification of components that need to be made survivable. In fact, it must be understood that survivability challenges extend beyond merely keeping nuclear attack assets such as warheads or their delivery systems safe from attack. Of course, the atomic bomb is at the heart of the matter and must survive for 'nuclear' retaliation to be mounted. In fact, as and when international commitments such as the FMCT prohibit quantitative additions to the nuclear stockpile, guarding the available warheads will become even more critical. But survivability of the bomb alone cannot suffice.

Credible deterrence demands the survivability of other enabling mechanisms and supporting structures too. In fact, the *nuclear weapon* or its *delivery mechanism* would mean little in the absence of an *alive and able decision-maker* at every level in a clearly defined chain of succession, a *command and control system* that provides relevant inputs to the decision-maker, and a *communication network* that carries the decision right down to the *man in the field* who is to execute the launch, besides providing him *accurate targeting coordinates* and other *supporting logistic elements*. Most importantly, inherent in this entire process is the survival of the *will to undertake retaliation*. This, in fact, is the most critical element because the others would be meaningless if the leadership of the moment is unable to bring itself to retaliate after a nuclear attack. But, the survival of this will is the most difficult to ensure, given its intangible nature. Of course, as is explained in the following section, certain specific measures can enhance the chances of survival of all components of the nuclear arsenal, including political will.

However, while mathematical modelling can help calculate the chances of survival of other components, nothing can guarantee, or even exactly assess, the survival of political will. Fortunately though, this applies equally to the adversary since he cannot calculate the response of the leadership with any certainty either. Pakistan, for instance, has miscalculated on this count in the past wars. Especially in the case of Kargil, it assumed that a caretaker government would not have the will or the gumption to take any decisive action against the Pakistani soldiers in the guise of *mujahideen*. A strong belief in the martial superiority of its own nationals has often prompted Islamabad to undertake military adventures that have gone awry. However, an ill-conceived nuclear misadventure would cost both the countries dear. Pakistan does not have the capability to undertake a disarming first strike against India and it must not assume that New Delhi would not retaliate because such an assumption would spell catastrophe for the region.

How to Ensure Survivability?

There are several ways by which to enhance the survivability of the various constituents of the nuclear arsenal. Every nation makes its choices based on different considerations. While information on the exact *modus operandi* of ensuring survivability would naturally be classified, it is important that the adversary be made well cognisant of the fact that steps are being taken to this effect. Communication of this resolve through the right kind of signalling is critical for enhancing deterrence.

Amassing a large stockpile of nuclear warheads or delivery systems is not in any way a guarantee of making them more survivable. Survivability requires, instead, a more intelligent approach that optimally mixes survival measures such as secrecy, deception, dispersion, concealment, mobility and defences. The determination of how much to conceal and where, or what to make mobile and how, and what to geographically disperse must be made on a considered assessment of the adversary's and own strengths and vulnerabilities. This co-relation will become clearer as several options of survivability, based on their costs and benefits, are examined in the following paragraphs.

Secrecy

Limiting access to information about the extent and location of nuclear attack assets by keeping low the number of people, and hiding their identity is one of the simpler and cheaper ways to ensure survivability. In fact, this tactic has been employed in every nuclear weapon state to build a level of opacity by not having too many people in the know and encouraging guesswork to foment perceptions without revealing, nor denying, actual facts on several matters nuclear.

In the case of India, the culture of secrecy has deep, historical roots, given that the 'wise ones' were never amenable to easily sharing their knowledge with others. The Brahminical culture, in fact, is based on monopolising knowledge in a few hands. The bureaucratic system developed by the British, and as it exists today, is also given to functioning with a high level of secrecy. Most scientific and defence organisations too work on a 'need to know' principle. While this mode of functioning can hamper the development of a more formal and institutionalised system for managing the nuclear deterrent, it nevertheless enhances its security because not everyone has sufficient information about the constitution, position, or disposition of the nuclear arsenal. This has its advantages, as was explained by Ashley Tellis[2],

> Since the entire organizational structure places a premium on extreme secrecy... potential adversary has to reckon with the prospect that there could always be some further strategic capabilities or technical resources held in reserve... unknown even to those few individuals otherwise thought to possess 'perfect' knowledge about the status and disposition of India's distributed strategic assets.

Thus, through a high level of secrecy where the number of people in the nuclear loop is deliberately limited, India seeks to deny its adversaries the information they would need to perfect their targeting strategies for a devastating first nuclear strike.

---

2. Ashley Tellis, *India's Emerging Nuclear Posture: Between Recessed Deterrent and Ready Arsenal* (Santa Monica: RAND, 2001), p. 422.

Deception

If secrecy is a passive measure to maximise survivability, deception is a more active method to deliberately mislead the adversary. This may be done through willful communication of false information or deliberate ambiguity through contradictory statements. For instance, Soviet President Khruschev contributed to the myth of a 'missile gap' in favour of the USSR when he made the statement that his country was producing missiles like sausages, even though the reality was very different. Besides verbal misinformation, deception may also be practised by building dummy missiles or launch and storage sites in order to multiply targets and reduce the adversary's chances of being able to hit all, or even all the correct targets.

While India has largely resorted to secrecy rather than deception in nuclear affairs, the production of dummy missiles or launch sites would be an effective and relatively cheaper way of ensuring survivability, especially of delivery vehicles. Presenting several targets to the adversary would sufficiently complicate his calculations to deter first use since he could never be sure that enough would not survive for retaliation.

Hardening

Shielding physical structures through use of special materials able to withstand nuclear attack is another way of ensuring survival of critical assets. It amounts to increasing the ability of structures, systems and components to tolerate exposure to the effects of nuclear detonation such as air blast, ground shock, electro-magnetic pulse (EMP), heat, pressure, and radiation. In an effective and widely prevalent use of this method during the 1960s and 1970s, the superpowers constructed hardened silos to keep their land-based nuclear forces (missiles and aircraft) safe from an attack. However, two technological developments have since reduced the efficacy of silos. Firstly, modern space-based systems have the capability to expose the position of silos for easy targeting; and secondly, the development of precision munitions and earth penetrating weapons has degraded the survival chances of nuclear assets in a silo.

In the case of India, however, silos have certain advantages and disadvantages. For instance, given that neither Pakistan nor China yet have adequate space-based capabilities for accurate targeting, or even very reliable accurate missiles or earth penetrating weapons, hardened silos or storage in deep caves or tunnels remains an option for the near future. But it must also be realised that Chinese missiles are rapidly moving towards greater accuracy through GPS enabled systems. Given their focus on enhancing national military space capabilities, their ability to target silos would improve dramatically in the coming years. Meanwhile, silos are expensive and difficult to build given the need for special materials and other considerations. It involves hardening not only the physical outer structure but also constructing the exact spaces for hosting nuclear assets in such a manner that even individual components can absorb violent ground motion. While the Department of Atomic Energy (DAE) and Defence Research and Development Organisation (DRDO), as custodians of nuclear weapons, are believed to already have specially constructed sites for storage of nuclear warheads, silos for delivery systems maintained with the military missile units would have to be specially constructed with suspension devices that can support ground motion and are made of materials hardened enough to withstand the effects of nuclear blast. Power supplies, communication and launch control electronic hardware of the delivery vehicle would also have to be protected against thermal effects, ionospheric disruptions and radiation effects.

The location of silos would also have to be carefully considered on the basis of the range of adversary missiles/aircraft as well as proximity to own launch sites for quick reconstitution of retaliatory forces. Even though the NFU strategy reduces pressure of immediate retaliation, unnecessary loss of time would not only raise the risks of another wave of nuclear strikes but could also adversely affect own resolve to counter-strike as international pressure to show restraint mounts and news on extent of damage flows in. The latter issue and its impact on decision-making is addressed in some detail in the section on preparation of resolve.

Besides silos for housing attack assets, one other major component of the nuclear arsenal that could be considered for placing in hardened structures in times of crisis is the National Command Authority (NCA) and its alternate in a hardened National Command Post (NCP). The NCA is the decision-making body comprising the Prime Minister and other Cabinet ministers who are tasked with the responsibility of authorising nuclear use. The NCP, meanwhile, is a robust communication centre with the ability to receive information and disseminate it. Gen Sundarji distinguished between the two as, "If NCA is the brain, the NCP is the nervous system including the sensory functions."[3] Obviously, the survival of both is essential for retaliation, and to ensure their continued existence and ability to function, these command, control and communication nodes must be hardened and deep underground. This facility would have equally hardened communication systems to other nodal points in the nuclear command chain. Obviously, technical, technological and financial complexities would be involved in constructing the facility and even more so in deeply burying the entire information distribution network over long distances. However, none of these challenges is insurmountable if the efficacy of such a structure is certain. The problem lies in the fact that as the accuracy and

3. Gen K. Sundarji, *The Blind Men of Hindoostan* (New Delhi: UBS Publishers Distribution Ltd, 1993), p. 89.

lethality of adversary missiles and weapons improve, along with their intelligence gathering capabilities, the advantage of such hardened structures would rapidly erode. Nevertheless, hardened command and control structures with the requisite redundancies in secure and robust communication links, offer one means of making them survivable. In fact, once adequately hardened structures of the NCP and its alternate have been built, these must be made transparent to the adversary to communicate the credibility and certainty of retaliation in the case of a nuclear attack.[4] Knowledge of the existence of these structures would not only provide confidence to own side, but also make the adversary conscious of his inability to execute decapitation in the hope of escaping punitive retaliation.

At the same time, depending only on hardened structures cannot be possible given their cost, complexity and vulnerabilities that would only increase over time. In the case of those components of the arsenal that can be easily made mobile, delivery vehicles, for instance, that can be road or rail mobile, deep buried, hardened structures need not be the preferred choice. Mobility, in their case, is another significant survival measure that must be carefully examined.

Mobility

One way of circumventing the vulnerability of nuclear assets in silos is to make them mobile. If these were frequently moved around on an elaborate road and rail network, it would be impossible for the

4. In the case of the USA, it is well known that the National Military Command Centre (NMCC)is situated under the Pentagon and the Alternate NMCC beneath Raven Rock mountain in Pennsylvania. The US also maintains an airborne command centre or the National Emergency Airborne Command Post. In the UK, the Primary Command Centre (PCC), codenamed Pindar, is located beneath the MoD building in Whitehall and is connected by underground tunnels to Downing Street. The alternate PCC is an underground complex at Hawthorn, near Bath.

adversary to constantly monitor and accurately target these forces. This, of course, would require making the nuclear assets smaller, missiles based on solid fuel and automated to the extent possible. If dummies were also added to the actual, mobile forces, it would further complicate the targeting requirements of the adversary and, thereby, enhance deterrence because he could never be sure of the numbers of real forces that would survive his first strike.

However, two parameters could constrain the extent of mobility: first, the ability to reconstitute forces quickly after attack. If the assets are too widely dispersed, it might prove to be logistically difficult to quickly bring them together for retaliatory launches. This, nevertheless, is not an insurmountable challenge and can be overcome through thorough pre-planning in peace-time and conduct of periodic simulation exercises to understand and overcome limitations; the second constraint on mobility is imposed by communication lines. Unless these are secure, hardened and sufficiently redundant, it could cripple the retaliatory system by the sheer inability of mobile units that have survived a first attack to link up with one another or the NCA. Therefore, degradation of communication systems could prove to be a particular point of vulnerability and adequate attention must be paid to make these survivable so that the benefits of mobility of forces can be maximised.

Besides nuclear attack assets, making the NCA/NCP mobile, either on an airborne platform or on land transportable vehicles is also worthy of consideration in the Indian context. With the acquisition of the airborne warning and control system (AWACS) platform and with aerial refuelling capabilities, India does have the capacity of making the command post airborne in crisis situations. This may not be dismissed as preposterous because unlike the immediate mental link with the American 24-hour air readiness, in the case of India, it would not be necessary to maintain such a facility on constant alert. It would only amount to configuring a command force that would be available for such functions as and when necessary and for occasional exercises to maintain the capability. Similarly, on land, India's extensive rail and road network offers an option of a safe haven for the command authority to function from. It could be intelligently and safely knitted into the larger civilian network, though there would be a requirement for specially constructed camouflaged vehicles (whether rail bogies or road carriers) that can cater for sufficient reserves of power to run data and communication systems, sufficient fuel for adequate movement, and other logistic requirements to ensure independence of movement. Air, rail or road mobility could offer relatively less expensive and more readily available options for providing survivability of the NCA/NCP. Pre-planning with adequate forethought can equip these options with greater redundancy at much lower costs and levels of complexity, thereby ensuring a sanctuary for the national leadership to survive an attack, assess damage, contemplate retaliatory options and order a counter-strike.

Dispersion

There are two ways of exercising dispersion of nuclear assets in order to ensure their survival. One of these is to geographically distribute capabilities/systems over several locations in such a way that no complete strategic systems exist as transparent targets. In fact, that is the state in which the Indian nuclear doctrine mandates the forces normally be maintained. Weapon cores, weapon assemblies, missiles, and their launch vehicles are all located at different sites, to be brought together as "fully employable forces" only in case of a crisis. This proffers the obvious advantage of multiplying targets to complicate adversary calculations. As was explained by Gen Sundarji, "It is not just a question of [finding] 'needles in haystacks' but parts of many needles in many haystacks which might be brought together when required within hours to days, to form full needles in yet many more different haystacks."[5] Of course, this kind of dispersion does pose the challenge of timely and effective reconstitution of the nuclear force during a crisis. It is normally assumed that any crisis between India and Pakistan or India and China would gradually develop over a period of time. Conventional hostilities would have reached a certain level of escalation before arriving at the point of possible nuclear use. Therefore, components of the nuclear force would have to be brought together in a period in which conventional strikes would be on. Movement of nuclear assets during such time would suffer from challenges of their own. It would call for elaborate planning and coordination among different agencies to remain networked to ensure safe passage on every mode of transportation likely to be used.

The second mode of dispersion is to spread the nuclear assets over a range of delivery platforms. Historically, every state with nuclear weapons has used air delivery as the first option because of its ready availability. However, given the restricted range of aircraft and their limited penetration capabilities in a dense air defence environment, missiles – land-based and sea-based – have evolved as the preferred option. Of course, air-launched, supersonic cruise missiles like the Brahmos and its follow-on systems offer a credible option.

While mobility is an important aspect of land-based missiles[6], the highest level of survivability is, nevertheless, believed to lie in placing nuclear tipped missiles with sufficient ranges on nuclear

---

5. Gen Sundarji, "Indian Nuclear Doctrine –I: Notions of Deterrence", as cited in Tellis, n. 2, p. 426.
6. Interestingly, the USA, given the luxury provided by its geography and the nature of its threat perceptions, has traditionally not had any land-mobile missiles in its arsenal.

powered submarines. This has been a time-tested formula and from the time the submarine launched ballistic missiles (SLBM) first became available after 1955, it has been considered the most survivable delivery system. A US nuclear submarine captain described the American Polaris submarine as,

> ....an extremely survivable assured capability that the Soviets knew they could not destroy and knew if they conducted a first strike, that system would some day be available to retaliate. It might take some time to get the message to them from a destroyed national headquarters, but, at some day, the missile warheads would come raining in and they would pay the price."[7]

Indeed, every NWS has aimed for a triad of nuclear forces and in countries that have over the years, in deference to their changing threat perceptions, given up some nuclear delivery platforms, have still maintained SLBMs for their high survivability quotient. For instance, the UK presently maintains its nuclear forces only on its four submarines and France too maintains a dyad in the SLBM and air delivery platforms.

In the case of India, the nuclear doctrine provides for the constitution of a triad. Given the security scenario in the neighbourhood, the eventual induction of the SLBMs would indeed provide a higher guarantee of survivability. However, sea-based deterrence poses its own challenges, two of which are the acquisition/development of the vessel itself and the development of necessary communication systems with the submarine. For India, the construction of the submarine has not proved easy given that the country has had to undertake a completely indigenous development of the vessel, having been denied import of nuclear and even dual use materials and technology.

---

7. Gerard J. DeGroot, *The Bomb: A History of Hell on Earth* (London: Pimlico, 2005), p. 267.

Also, since the country's nuclear energy programme has been based on the development of Pressurised Heavy Water Reactors instead of Pressurised Water Reactors, the technology best suited for nuclear submarines, the country has had to develop a parallel track of technology. The second challenge arises in the form of establishing secure, constant and reliable channels of communication with nuclear submarines. Normally, underwater communications are possible through very low frequency (VLF) bands of the electromagnetic spectrum. These communication channels, however, have a restricted traffic bearing capacity and are slow. Also, their transmitting stations are large, fixed and difficult to harden, making them vulnerable to a first strike. The US resolved this problem by having an airborne VLF system coupled with satellite communications or by developing extremely low frequency (ELF) communications. India will have to find its own answers to this problem.

There are a few issues that must be examined with regard to sea-based deterrence. Firstly, given the large Indian landmass and the gigantic inland road and rail network that could be effectively used for mobile missiles with adequate ranges, could their survivability not be ensured on land any better than it could be at sea? In a situation of fast improving anti-submarine warfare (ASW) capabilities, are SSBNs that would be carrying concentrated clusters of strategic capabilities (at least 12-16 MIRVed missiles equalling 96 warheads) more or less risk prone? Of course, SSBNs are also most vulnerable when in port since they are difficult to hide. Moreover, given that there are not too many Indian ports that could host SSBNs, their targeting by the adversary could be relatively simple. Above all, the seabased leg of the triad does pose challenges of command, control, and communication, as well as those of pre-delegation of authority – requirements not catered for in the present command and control system, which is assertive in nature. Unlike land-based nuclear capabilities that can be maintained in a distributed form, a sea-based deterrent presupposes complete systems on board at sea. Once this leg of the triad becomes meaningfully operational[8], which should be some time in a decade or so, it would call for the development of technological and organisational arrangements to cater for chances of an accidental or unauthorised launch of a nuclear weapon from the sea. Even more than operational issues, once an SSBN force is ready, India will have to take the critical decision of graduating from the present posture where the civilian leadership exercises complete control over nuclear assets to one wherein custody of a number of nuclear weapons would reside with uniformed personnel. This transition from what Ashley Tellis describes as "force in being" to a "ready arsenal" would bring its own sets of implications for India's nuclear strategy and civil-military relations.

However, notwithstanding the above-mentioned problem areas of sea-based deterrence, it still offers enough advantages as a viable and effective option for enhancing survivability that does not allow it to be dismissed lightly. The mere fact that all NWS perceive greatest survivability in this leg of the triad is not without reason. Indeed, for a peninsular nation like India, the vast seas around it do provide large areas where SSBNs could remain hidden with a significant nuclear arsenal for long periods of time to mount retaliation if and when necessary. In fact, the credibility of a counter-strike is ensured once an adversary knows that a fully armed SSBN is out at sea. This reinforces the certainty of retaliation by making counter-strike almost automatic, in case the country has been attacked with nuclear weapons.

As far as other issues such as vulnerability in port or problems of C3 are concerned, these are not insurmountable challenges

8. The reference here is to a set of SSBNs and not just one of them.

and the Indian Navy is engaged in resolving them. For instance, it is planned to have three SSBNs by 2015, of which one would be kept in reserve while two remain out at sea by rotation[9], thereby reducing the chances of their being caught in port. Similarly, the possibility of unauthorised use of the weapon is resolved through electronic locks on weapons that would need to be operated by more than one person. In order to further ensure the survivability of SSBNs, some other precautions will also have to be worked at. The first of these is rigorous testing of SLBMs with adequate ranges that would enable SSBNs to stay out of harm's way. As of now, the Indian advanced technology vehicle (ATV) would be equipped with a solid fuelled 750 km range SLBM. But the DRDO has plans to equip the SSBNs with the extended range Agni III of 5,000 km. Secondly, development of adequate anti-ASW capabilities would be critical. Making the submarine as silent as possible and equipping it with some stealth features would certainly help, and efforts need to continue in these directions. At the same time, particular attention needs to be paid towards hardening the shore-based communication centres of the SSBNs because these are points of vulnerability.

Finally, it may be said that despite some vulnerabilities, sea-based deterrence certainly has the greatest chance of being survivable and providing the most credible deterrence through the right kind of power projection. And since deterrence is essentially a mind game, India will have to invest in some minimum level of sea-based nuclear capability as part of its credible minimum deterrence.

However, till such time as the sea-based deterrent becomes available, India would have to base its nuclear deterrence and manage its survivability challenges with land-based ballistic

---

9. Sandeep Unnithan, "The Secret Undersea Weapon," *India Today*, January 28, 2008, p.52.

missiles (by making them more solid fuelled and more rail and road mobile) and air delivered nuclear warheads (by hardening air bases and equipping them with effective air defences). Of these two, in the immediate future, the responsibility of India's nuclear deterrence will have to be borne mainly by the air force, since ballistic missiles with the required ranges and accuracy are yet not operationally available. In order to ensure flexibility of selection of missile launch sites for enhancing survivability, it would be necessary for India to develop missiles of the ranges of not less than 5,000 km. Even then, air delivered nuclear capability will play a critical role in deterrence since manned nuclear capable aircraft provide an effective means to show strength and resolve and yet give the political leadership the flexibility of recalling a mission even after it has been launched.

Active Defences

The deployment of air and missile defences around critical points is another way to ensure survivability by intercepting incoming missiles or air attacks before they hit the target. Limited missile defences were used during the Cold War years for increasing survivability of land-based assets. However, to maintain vulnerability for mutual assured destruction, BMD was allowed only on a limited number of sites. After the abrogation of the ABM Treaty, this restriction has been lifted and the US is now engaged in deploying a multi-layered highly advanced national missile defence to eventually provide protection to the entire American landmass. This is envisaged through the establishment of an elaborate network of radars and interceptors at different sites within the country and outside. Obviously, such a system is technologically challenging as well as politically destabilising.

While these issues are dealt with separately in the chapter on BMD, the consideration of missile defence for India, from the perspective of survivability, needs to be examined here. Theoretically, of course, there can be no denying that a system that can intercept incoming enemy missiles and neutralise them before they hit the target ensures the survival of what they are meant to protect. However, there can be little guarantee that every incoming missile will be intercepted in time. The financial and technological costs and complexities are not of any small dimension. These, in fact, are exacerbated by the more demanding geographical constraints and the more advanced capabilities with adversaries that an Indian BMD system must cater for in contrast to the situation of the USA. In fact, simple counter-measures can defeat BMD and it is relevant to point out that in the Indian case, this is especially important since China, over the last decade, has concentrated on developing effective counter-measures to defeat a far more sophisticated BMD of the US.

Over the last couple of years, the DRDO has conducted some successful interceptions that have raised interest and confidence in the technology. Intelligently complemented with some imported systems, BMD could offer some enhancement of survivability for deterrence. However, its utility must be carefully tailored to the Indian security environment. For example, since survivability of retaliatory forces is a prerequisite for assured retaliation, erecting point or area defences over some types of nuclear assets such as over early warning systems, air bases for nuclear capable aircraft, command posts, submarine communication centres, nuclear production facilities, launch or storage sites in cases where mobility is not enabled, must be considered. But it would be unnecessary as well as unfeasible to opt for missile defences over many cities. Of course, erection of BMD over critical points might suffer from the disadvantage of exposing their location to the adversary and, thus, subverting the advantage of concealment. Yet, at the same time, erection of defences also injects uncertainty into the mind of the adversary and complicates his targeting calculations, and can be used to that extent.

Preparation of Resolve

Raising awareness and exposing decision-makers to simulated exercises in which escalation to the nuclear level is envisaged could be some of the preparatory tasks towards ensuring survival of the will for retaliation. As has been mentioned earlier, political will is the least tangible component of the nuclear arsenal and there can be no prediction on how it would react during nuclear war. For instance, the news of a very high level of damage could affect the decision-maker in two ways: on the one hand, it could send him into a state of shock and lead to action paralysis; on the other, it could also lead to immense anger and immediate action. Low damage of nuclear attack could also make the decision-maker more susceptible to external pressures and constrain the scope of action. The location of attack could also influence the mental frame of the decision-maker. An isolated nuclear attack on an air base, or a surface ship out at sea, or in a remote desert army unit would, in all certainty, affect the decision-maker differently from a situation in which the adversary has mounted multiple counter-force attacks coupled with some counter-value ones too. The Indian nuclear doctrine clearly mandates massive retaliation in every case of nuclear use . But would the leadership be able to make the difficult decision to arrive quickly at a response that is equally able to meet the requirements of domestic expectations, international pressures and, at the same time, most importantly, send the necessary message to the adversary. The last criterion, in fact, is critical because this response would determine the credibility of deterrence for the future. Of course, the nature of response would have immediate implications for the country, but it would also influence perceptions for the future. In this context, it becomes extremely important for the political leaders to adequately understand and appreciate the intricacies of nuclear deterrence since they are the prime decision-makers in the Indian system and also need to convey credible nuclear signals in peace-time and war.

It is obvious that survivability is achievable through a number of measures. The challenge lies in making the right choices based on the most relevant parameters. The first of these must be an assessment of the adversary's intelligence, reconnaissance, surveillance, target acquisition and strike capabilities. For instance, in order to evade better human intelligence skills of the adversary, it would be necessary to maintain a high level of secrecy on information of assets and their locations, capabilities etc. Compartmentalisation of information within the government, armed forces and even strategic organisations would be necessary, besides elaborate and sophisticated personnel reliability programmes in every establishment and at every level. On the other hand, a higher adversarial capability of peeping into own territory through surveillance would entail greater emphasis on deception and mobility. Or, the capability of adversary to conduct effective electronic warfare would imply placing greater emphasis on making own communication networks more secure and redundant.

How Much to Make Survivable?

In any kind of a counter-force, first strike mounted by the adversary, a certain amount of attrition of the nuclear forces should be expected. However, the key to credible deterrence through punitive retaliation lies in understanding, and arriving at, an approximate figure of own nuclear assets that *must survive*. This, in turn, has to be a function of a considered assessment of how much would be required to impose an unacceptable level of punishment on the adversary. At least that much has to be made absolutely survivable. However, the assessment of the damage threshold of the adversary is a complex calculation. During the days of superpower rivalry, the US had arrived at a complicated number: according to McNamara's calculation, the US had to destroy 50 per cent of the Soviet population, and 25 per cent of its military and industry for it to be considered unacceptable by the USSR. Seen in retrospect, this is today considered a horrendous over-assessment. But one can never assert with any certainty how the adversary would make his calculations of unacceptability. Mao Tse Tung, for instance, had suggested a very high Chinese tolerance threshold, ready to brook the loss of even 300 million countrymen, because as he pointed out, even then, enough Chinese people would survive to keep China going. With modern levels of development, however, it is assumed that countries would have lower damage tolerance thresholds. Also, greater role of public opinion enabled by an explosion in information and the media clearly indicates a weaker stomach for damage to life and property. However, in the case of India, three essential parameters, as described below, can be used to assess how much must be made absolutely survivable for assured retaliation.

Assessment of Unacceptable Damage

The first parameter that can help assess how much must definitely survive with India is the calculation of what would constitute 'unacceptable damage' for the adversary. Therefore, the stakes must be understood and considered from the point of view of the adversary. While this cannot be an easy task, certain educated guesstimates can be made on the basis of following factors.

Socio-Political System of the Country

This would have a deep impact on how decisions are made by

nations. A high level of democratic openness of society, a large number of stakeholders in decision-making, and freedom of the media can cast a constraining influence on an adventurous political leadership. In contrast, a system that is autocratic, shows little respect for public opinion, restricts and controls information dissemination and does not care for legitimacy of action would be more prone to taking nuclear risks. In the latter case, the level of damage that the nation would be willing to absorb would be much higher than in the case of the former. This is amply proven by the fact that China, even with a small nuclear arsenal, is able to effectively deter the thousands of weapons in the American nuclear armoury. This is because the socio-political system of the US cannot sustain damage to itself. It has a low damage tolerance threshold, low enough for China to threaten even with its relatively modest capabilities. For India, however, the situation is the exact opposite. Its deterrence has to be imposed upon a nation (China) that has a high limit of damage acceptability, in which the decision-making is confined to a very small number of leaders, and where the media is highly controlled. Therefore, the size of its arsenal must take these factors into account.

Strategic Culture

The propensity for undertaking and absorbing military casualties is also influenced by the overall strategic culture of the nation. This, in turn, is significantly influenced by the country's historical experiences of war. A country that has a self-image of having been wronged in a war or having been deprived of something perceived as its own would be more tolerant to bearing damage or suffering 'costs' for righting the past action. Also, a revisionist power is more damage tolerant compared to a status quo nation. In the case of India, Pakistan and China are both revisionist and, hence, logically should be expected to be ready to bear more damage.

Issue at Stake

The level of acceptability of damage is also influenced by the amount of value a nation places upon the issue at stake. Of course, it could well be argued that nothing can be valuable enough to merit the kind of destruction that a nuclear exchange would bring. By this logic then, the possession of Kashmir could not be worthwhile for Pakistan if it meant the loss of the rest of the nation, especially the Punjab. Or, that China would not want to lose Beijing in exchange for gaining Arunachal Pradesh. However, nuclear use would seem acceptable in case the country was pushed into a corner with its survival at stake and compelled to nuclear use as a measure of last resort.

Level of Economic Development

An economically more developed country is normally expected to have a lower damage tolerance threshold. This is because an impoverished country in any case has less to lose and is, therefore, willing to accept more damage. It is for this reason that it is assumed that China of today has more to lose and, hence, would be less willing to suffer damage to the level of development that it has got habituated to. As the level of economic development increases, nations become more attached to a certain way and quality of life and, hence, develop a weaker tolerance level for any loss of their comfort zone. Those that have nothing to lose are more acceptable of even higher levels of damage.

Reliability of Own Arsenal

The second parameter on the basis of which the quantum of arsenal that must survive can be assessed is the reliability of own nuclear warheads and delivery vehicles. The higher the reliability that the missile would be able to carry the nuclear weapon to the desired target, and that the weapon would explode to provide the expected yield, the less is the amount that needs to survive. More reliable systems can infuse greater confidence that whatever survives would be able to do the necessary damage and there would be less requirement for building redundancies into calculation of numbers. However, reliability needs to be established at various levels. For instance, missile reliability should imply the ready availability of a deliverable missile at a given moment in time. It cannot include missiles that are under repair or maintenance or not ready for immediate action. Communication reliability would have to stem from the quick and efficient dissemination of orders to launch. Launch reliability would require that the missile lifts off when so ordered. Booster reliability would entail the igniting of the rocket in time to send off the missile. Also, having done that, the boosters must also ensure separation reliability so that they can disengage from the missile in flight instead of hanging on to it and interfering with its ballistic flight. Thereafter, penetration reliability must assure that the missile would actually be able to home in on the pre-determined target. Lastly, detonation reliability would imply the absolute certainty that the warhead explodes over the target and provides the expected yield to inflict the kind of damage that was considered necessary. Therefore, through a considered calculation of these combined individual reliabilities, it could be established as to how much a nation must make survivable.

Purpose of Own Nuclear Weapons

The last essential parameter that can be used to decide how much must survive the adversary's first use of nuclear weapons would be the purpose of one's own nuclear retaliation. Is it to wreak punishment on the first user for his act? Or, is it to bring about total destruction of the adversary as a form of revenge? In the case of the former, the survivability requirements can be less stringent than in the case of the latter. Imposing punishment could be possible with the use of a few fission weapons in counter-value mode. Of course, this is affected by the adversary's determination of how it perceives its own damage. But, in the case of most modern societies, and given the densities of populations in the region that India inhabits, even a score of nuclear weapons could bring untold destruction of life and property. On the other hand, if the purpose of the nuclear arsenal is to completely decimate the first user, then obviously, more would need to be made survivable.

Given the purpose of the Indian nuclear weapon, as established in the country's nuclear doctrine, it exists to impose deterrence. As has been reiterated several times through the chapters of this book, the nuclear weapon is perceived as a political tool to ward off nuclear blackmail or coercion. Therefore, India does not visualise any situation in which the weapon could actually be used for war-fighting. However, in order to keep it that way, it becomes important to make the necessary arrangements and convey the same to the adversary that the country is ready for any kind of deterrence breakdown. Ensuring the survivability of the nuclear arsenal, therefore, in this scheme of things is the most critical basis for establishing and sustaining credible deterrence.

7

Lure of the Shield

BMD in India's Nuclear Strategy

Between 2006-08, the march of India's indigenous technology towards the development of anti-ballistic missile (ABM) capabilities attracted much attention. According to officially released information, on November 27, 2006, in an exo-atmospheric test 50 km above the earth, a Prithvi ballistic missile was successfully intercepted by a second modified Prithvi interceptor missile. Then, on December 6, 2007, India fired a hypersonic interceptor missile that destroyed an incoming missile in a direct hit over the Bay of Bengal. This endo-atmospheric interception took place at an altitude of 15 km. These tests revealed that the country was working on acquiring some sort of ballistic missile defence (BMD) capabilities. The Chief Controller of the Defence Research and Development Organisation (DRDO), V.K. Saraswat, claimed that his organisation was "developing a robust anti-missile defence system that will have high speed interceptions for engaging ballistic missiles in 5,000 km class and above."[1] Indeed, the exo- and endo-

1. Animesh Roul, "India: Missile Defence Dreams," *ISN Security Watch*, March 28, 2008.

atmospheric interception tests were major breakthroughs. A third trial interception to achieve a higher kill altitude (between 50-75 kms above the earth) with a modified Prithvi equipped with inertial navigation to perform manoeuvres to engage the incoming missile is next in the line. V.K. Saraswat claims, "With the third test, we will have the entire BMD capability to detect, intercept and destroy intermediate range and intercontinental ballistic missiles in the 5,000 km range coming from any country."[2] The system is expected to be ready for operational deployment by 2010-11. While these dates do look a bit over ambitious, the claim too actually only means that by the start of the next decade, India would have developed a deployable system capable of intercepting enemy missiles. Its actual integration into the security architecture would obviously take time, as it would span a number of activities and deal with a range of conceptual and operational issues.[3]

For the moment, in fact, India's technological developments appear to have gone ahead of the conceptual issues of nuclear strategy related with the induction of BMD. A survey of the writings on BMD by Indian strategic analysts or official statements by political or military leaders provides no clear indication of a considered, resolute move towards the development and deployment of BMD. Of course, it is well known that the Indian government in 2001 had rather uncharacteristically welcomed the US move to supplement its nuclear deterrence with defence, but

---

2. "India to Conduct Third Interceptor Missile Trial", www.thaindia.com, August 29, 2008.
3. The situation may be compared to the one obtaining in Russia as regards its S-400 ABM system. Its operational readiness for deployment was announced in August 2007, but it is estimated that it will take upto seven years before the Russian armed forces can field 24 battalions with the new systems because of the practical reality of the industry being able to produce only a limited number every year.

this approach became more cautious within the next few days.[4] Since then, there have been talks between the US and Indian administrations on the subject, but no major public debate on the pros and cons of BMD for India has taken place. Nevertheless, given the determined and rapid manner in which the DRDO is pursuing the technology, it appears a given that India has decided to develop and deploy some form of BMD. Even so, several questions must be answered in order to make optimum use of these capabilities while also understanding their limitations.

What exactly is India looking for? How good will an Indian BMD have to be to meet the threats the country faces? Is such a system technologically feasible, financially affordable and politically possible? While R & D continues to overcome technical hurdles, it is critical that correct political choices are made after understanding the conceptual underpinnings of the BMD for India's nuclear strategy. Clearly, missile defence has certain benefits. But, it has limitations and far-reaching implications too. Therefore, a clear-headed assessment of *what* BMD can achieve for India, and *how*, is imperative. This would not only help provide direction to the ongoing technological efforts, but also ensure optimal utilisation of the limited resources available with the country.

It is in view of the above that the first section of the chapter explores the conceptual relevance of BMD for India in the context of its nuclear deterrence. It explains the relationship between deterrence and defence, which during the Cold War period was considered mutually exclusive, but is being conceptualised differently today by the USA. How does this apply to India's

4. The initial reaction of the Indian government to the May 1, 2001 speech of President Bush, in which he explained the logic of defences along with cutbacks in nuclear arsenal, was enthusiastic. But by the time then US Deputy Secretary of State Richard Armitage came to India within a few days of the speech at the NDU, with the stated objective of explaining the broad contours of the BMD, India had shifted to a more cautious approach.

nuclear strategy? The second section examines and compares the American logic for BMD with India's threat perceptions in order to assess how far BMD could be effective given the uniqueness of the Indian security imperatives. The last section finally extrapolates the nature and possible functioning of an Indian BMD and flags some of the operational issues that will have to be grappled with.

Relationship Between Deterrence and Defence

Having conducted five nuclear tests in May 1998, India had just set out on the path of establishing credible minimum deterrence when the US announced its intention of pursuing missile defence as a complement of deterrence.[5] In a landmark speech delivered at the National Defence University (NDU) on May 1, 2001, President Bush categorically stated that "deterrence can no longer be based solely on the threat of nuclear retaliation" because the world had become "a less certain, a less predictable one" in which missile defence was required to handle missile threats from "rogue states".[6] This conclusion echoed the sentiment voiced in the report submitted by a bipartisan commission instituted under the chairmanship of Donald Rumsfeld in the second half of the 1990s. It had concluded that the threat posed by emerging missile technologies was "broader, more mature and evolving more rapidly than has been reported in estimates and reports by the Intelligence community."[7] This provided the basis for the US

5. It may be recalled that attempts in this direction had been made from the 1980s onwards when President Reagan had proposed the Star Wars programme. But it was with President Bush Jr. that BMD acquired a definite purpose and direction.
6. President Bush's speech on missile defence, delivered at the NDU, Washington, May1, 2001, and as available on http://www.acronym.org.uk
7. Executive summary of the Report of the Commission to Assess the Ballistic Missile Threat to the US, July 15, 1998, p.1. As available on http://www.fas.org. It may be recalled that during the mid-1990s, the US National Intelligence Estimates (NIE) had downplayed any new missile threat to the

Administration to include the development and deployment of BMD as a major plank of its new national security strategy. Indeed, the Nuclear Posture Review of 2001 had also recommended a new security architecture premised on a novel triad of which defence (including BMD) was on 214e of the legs.

Of course, every country has the right to undertake periodic threat assessments and revise the measures it considers appropriate to defend itself. The US did the same when it concluded that BMD was essential for its security in view of the altered threats. But what made this decision controversial was the impact it was likely to have on the nuclear deterrence relationship that it had with the established nuclear powers, notably Russia and China. During the years of the Cold War, offence was deemed as the best form of defence. Mutual vulnerability was considered essential to hold nuclear deterrence in place because it was the near certainty of MAD that was believed to keep the use of nuclear weapons at bay. Accordingly, deterring nuclear attack by promising assured nuclear retaliation to cause unacceptable damage was the overarching objective of nuclear strategy and preparations to defend nations in the event of a nuclear attack were de-emphasised. Rather, it was argued, and largely accepted in Moscow and Washington, that defences could upset the deterrence equation that was otherwise "stable" when based on mutual vulnerability to nuclear retaliation.

Predominant wisdom on missile defence during the Cold War years, therefore, was based on two broad assumptions. One, that a BMD that could blunt a nuclear attack, would cast a destabilising influence on strategic balance by increasing the risk of war and eroding one's faith in one's own nuclear deterrent. In other

---

country. In 1995, the NIE had noted that North Korea was unlikely to obtain the technological capability to develop a long range ICBM and that there was an even lesser chance of Iran or Iraq developing such a capability.

words, it carried the risk of one country triggering off a nuclear war because it felt secure under its own missile cover and not afraid of retaliation. Two, it was presumed that the deployment of defensive systems would trigger an arms race for the upgradation of offensive capabilities in a bid to attack the other's perceived sense of invulnerability.[8] Besides these two assumptions, of course, the very technological feasibility of BMD was also in doubt, especially since the threat to be catered for was not a country with a few missiles, but the USSR that had several thousands of nuclear warheads and a formidable delivery capability. So, *confidence in stability of nuclear deterrence* was supplemented by *scepticism over the feasibility of defence.* Not surprisingly, therefore, Washington and Moscow preferred deterrence to defence.

It was in the context of these parameters that the US and the erstwhile USSR had signed the ABM Treaty in 1972. It prohibited the deployment of strategically significant BMD that could erode the other's nuclear deterrent. Both sides agreed to remain mutually vulnerable to each other's nuclear weaponry and hoped that this would prevent a nuclear exchange. Of course, Article III of the treaty permitted each side to deploy an ABM system at two sites (later amended to one site in 1974), each limited to a 100 interceptors. During the years of the Cold War that were marked by a lack of trust on either side about the other's intentions, this treaty did serve its purpose. With the end of bloc politics, however, and the emergence of new missile threats from smaller states, as well as rapid technological advances by the beginning of the first decade of the new millennium, BMD had acquired a new justification, and, hence, a new momentum.

Theoretically, BMD deployment changes the nuclear equation between nations by adding a definite denial dimension to nuclear

8. For more on both these viewpoints, see Keith B. Payne, "Looming Security Threats: The Case for NMD," *Orbis*, Spring 2000, p. 189.

deterrence that may otherwise be based only on deterrence by punishment.[9] Once BMD is introduced, nuclear deterrence comes to be exercised through both punishment and denial. It may be recalled that in the 1990s, the USA and Russia had come to agree that nuclear weapons were not for war-fighting since a nuclear war was unwinnable. But the presence of missile defence makes it possible to perceive nuclear weapons as usable weapons of war. A country secure under its missile defence may be tempted towards nuclear preemption based on the belief that having taken out most of the enemy nuclear arsenal in a first strike, any leftovers could be intercepted by its BMD. Therefore, there is no fear of MAD and, hence, at least in theory, nuclear war becomes winnable, especially by the nation that would use the nuclear weapon first. While the US is responsible for having set this change into motion, Russia and China, the two nations that perceive missile defence as a threat to their deterrence capabilities are compelled to follow suit in order to restore what they think is a disturbed equilibrium.

Six years after the abrogation of the ABM Treaty, the US and Russia disagree over BMD.[10] Russia is suspicious of

9. "Deterrence by punishment" is a theoretical construct of nuclear strategy attributed to Bernard Brodie. Explained simply, it underpins deterrence on the premise that "if you attack me, I'll punish you." On the other hand, "deterrence by denial" was extrapolated by W.L. Borden and may be loosely explained as "if you attack me, I'll hit back and you will lose the war." Therefore, while the former concept seeks to avert nuclear use by threatening punishment for the act, the latter imposes deterrence by implying the ability to win a nuclear exchange and thereby deny victory to the adversary.

10. It is another matter that not everyone within the US establishment or amongst the American strategic community is absolutely convinced about the desirability of BMD. The likes of Henry Kissinger have argued in favour of missile defence, "I never heard of a nation whose policy it is to keep itself vulnerable to attack." But others like Richard Garvin or Ted Postol question the feasibility of BMD. In 2007-08, the Congress denied the large sums of funding demanded by the Pentagon for its many BMD programmes. The Bush Administration has spent an estimated $60 billion on deployment of the BMD since 2002. "Missile Defense System Aimed at Potential Threats", www.npr.com, September 22, 2008.

American intentions, and fears that the planned missile defence deployments, especially in Europe (ten interceptors in Poland and a tracking radar in Czechoslovakia), could be easily expanded or reconfigured to target launches from Russia. While there may be no political reason evident in the present landscape for a nuclear exchange between Washington and Moscow, and the expansion or reconfiguration would have to be significantly substantive to threaten Moscow, nuclear deterrence is still held as the bedrock of national security and, hence, any move that upsets the deterrence-defence equation is considered threatening. It is hardly surprising, therefore, that Russia has been working over the last half a decade on enhancing the mobility and MIRVing capabilities of its ICBMs, the land-based Topol-M, RS-24 ICBM and, sea-based, solid fuelled Bulava-M to counter American missile defence (MD) endeavours.[11] Russia maintains that the most cost-effective and practical way of dealing with the BMD would be to equip its Topol-M with asymmetric capabilities such as "enhanced stealth, variability of flight trajectory and use of warheads capable of penetrating any missile shield."[12] More recently, the Russian supersonic nuclear bomber flew across the Arctic to Venezuela in a clear show of the reminder of the strategic prowess of Russian bombers armed with stand-off nuclear capable X-555 cruise missiles with a range of 2,000 miles that could easily bypass the US BMD.[13] That these

11. At the political level, Moscow had declared on July 14, 2007, that it was suspending its adherence to the Conventional Forces in Europe (CFE) Treaty and has threatened to withdraw from the Intermediate Nuclear Forces Treaty (INF). START II had already been declared dead when the US pulled out of the ABM in June 2002 and negotiations on START III have proved to be a non-starter.

12. A Russian military colonel cited in "Russia Test-fires Topol", United Press International, September 3, 2008.

13. "Tu-160s Tilt Nuclear Balance," *United Press International,* September 24, 2008. Russia is also planning to resume production of its new strategic bomber at the Kazank aircraft plant, "Russia's Nuclear Plans", *United Press International,* November 23, 2007.

would have to contend with US AWACS and F-15 is another matter. The point to note here is that while the US seeks defence against "unacceptable damage and assured destruction," Russia is honing its deterrence through enhanced offensive capability that could negate the defences.

For India, nuclear deterrence is the primary, and some would say, the sole purpose of its nuclear weapons. The deterrent relationship between India and its nuclear armed adversaries is based on the premise of, if not MAD, then at least the notion of each being able to inflict enough damage on the other that would be considered unacceptable and, hence, not worth a nuclear exchange. However, the introduction of BMD has the potential of injecting a certain loss of vulnerability and, hence, tends to upset the deterrence relationship, especially for the country with the smaller nuclear arsenal. Not surprisingly, therefore, just as China has been concerned with the implications of the US BMD on its nuclear deterrence, Pakistan has expressed concern at any BMD deployment by India.

However, there is a huge difference between the requirements, rationale, logic and, hence, implications of an Indian BMD as compared to the American missile defence. For one, the sheer scale of the two efforts is absolutely distinct. The US plans to deploy a multi-layered defence capability that would include ground, sea, airborne and space-based components in the form of tracking radars and interceptors to counter missiles in all phases of their flight – boost, mid-course or terminal.[14] The Indian attempts

---

14. The present US Global Missile Defence (GMD) consists of three echelons. The first is ground-based and includes two sites— Alaska and California— where ground-based interceptors (GBIs) are placed, supported by radars in Norway and Greenland. Poland has been identified as the third area for GBIs and Czechoslovakia for the radar. The second echelon comprises of the sea- based Standard Missile (SM)-3 on the US Navy's three cruisers and six destroyers, equipped with the Aegis information and control system. The third echelon, consisting of the detection satellites in space, supports the two layers.

in this direction would at best be for point defence through interception at the terminal phases. Secondly, unlike the US, BMD in India would be acquired by a nation that has a declared NFU doctrine. Therefore, one reason for New Delhi to acquire missile defence capabilities is based on the need to ensure survivability of critical components of the nuclear arsenal. Though survivability can be achieved through several other means, as discussed in the previous chapter, missile defence could be especially useful for defending command and control nodal centres such as the National Command Post (NCP) and the Alternate NCP. India's limited missile defence should, in fact, enhance nuclear deterrence and should logically be less disconcerting for the adversary since despite the BMD, an NFU posture reduces the likelihood of nuclear preemption, which is a primary concern when nations with a first use doctrine, like the US, deploy the BMD. Therefore, the deterrence-defence relationship in the case of India pans out quite differently from the US situation.

Comparison of the American and Indian BMD Logic

A Stratfor assessment traces the evolution of the US rationale for BMD in the following manner[15]:

> In the 1980s, it was about breaking out of the MAD impasse and wringing concessions out of the Soviets. Today, BMD has the potential to be something that was never seriously considered in the 1980s, a viable defensive weapon... BMD once was wielded as a political tool to *avoid* a future war; now it is coming to be viewed as a defensive weapon to be *used* in future conflict.

Three main developments were cited by US strategic experts and the Bush Administration as providing reason enough by

15. Peter Zeihan, "The New Logic for BMD", *Stratfo*r, March 6, 2007.

the end of the 1990s for a rethink on the ABM Treaty: end of superpower confrontation; emergence of new threats from smaller arsenals and the belief that these could not be deterred; and advancements in BMD technologies. The end of the period of hostility between the US and USSR and the gradual blossoming of a cooperative relationship timed with the emergence of WMD threats from smaller arsenals in states that the US perceived as being of proliferation concern. As the Rumsfeld Commission concluded, "A new strategic environment now gives emerging ballistic missile powers the capacity, through a combination of domestic development and foreign assistance, to acquire the means to strike (the US) within about five years of a decision to acquire such a capability (ten years in the case of Iraq)."

It was surmised that BMD could be effectively deployed against these countries that could otherwise not be deterred by threat of punishment. Lt Gen Ronald Kadish, director of the Ballistic Missile Defence Organisation (BMDO), now the Missile Defence Agency emphasised the inability of superior US conventional and nuclear weapons to deter countries that are bent on harming/destroying US citizens and assets. Citing from the Cuban missile crisis, he underlined that though the US nuclear weapons were able to deter the Soviets, they did not have the same effect on Fidel Castro who was "willing to sacrifice his country and 'die beautifully' in the process."[16] US designated rogue states are believed to harbour the same mentality and are perceived to be undeterred by offensive military might. But, BMD, it was felt, could be effective against these threats since the prospective missile arsenals of these countries were likely to be modest, in low, two-digit numbers that could be neutralised. Naturally then, the US Administration saw sense in

16. "US Missile Defense Director Cites Progress, Challenges," *Washington File*, US State Department, March 12, 2001. This article was adapted from a speech by Ronald Kadish to the Greater Fairbanks Chamber of Commerce, Alaska, on March 2, 2001.

deploying new and more advanced BMD technology to address its transforming threat perceptions. It was expected to safeguard American freedom of action in the face of coercive missile threats from regional powers, while also providing protection in the case of deterrence failure.

Consequently, in the post-Cold War world, missile defence reemerged in a transformed strategic environment, with a new mission, against a new kind of a threat, and equipped with newer advancements in technology. Over the last half a decade, MDA has conducted tests of several elements of the system and gradually confidence within the country and among US allies has grown in its capabilities. In any case, contemporary BMD plans are much less ambitious compared to what President Reagan had envisaged in the Strategic Defence Initiative (SDI) in the 1980s. It is not the purpose of this chapter to recount American successes and failures in the development of its missile defence capabilities. Enough literature is available to that effect. For the purpose of this study, these aspects are considered only to the extent that they are relevant for India's approach to BMD.

India's security concerns primarily emanate from its nuclearised neighbourhood. In fact, India happens to be in a region where nuclear proliferation has been an acute and worrisome reality for several decades now. China has been in possession of a fairly sophisticated nuclear arsenal for the past several decades and is resolutely modernising its conventional and strategic capabilities. For India, this is a source of insecurity, especially given the fact that China lays claim to thousands of square kilometres of Indian territory, and border disputes persist. Moreover, India has always been conscious of the growing imbalance of power in Asia due to the rise of an increasingly politically vocal, economically powerful and militarily assertive China.

Until 1998, when India finally decided to exercise its nuclear option in favour of weaponisation, the India-China relationship had existed as one very distant from MAD. China's nuclear preponderance was obvious. In fact, the threat from China was further compounded by its lack of compunctions about transferring its nuclear weaponry, designs and technology to other states, especially Pakistan, for commercial and strategic gains. Given the turbulent and traumatic history of partition and the continuing border disputes, India-Pak relations have never been smooth. In an astute move, China sought to exploit this situation by setting up a thriving nuclear and missile nexus with Pakistan to not only dilute US influence over Pakistan[17] but also strategically complicate India's security calculus. Therefore, from the 1960s to the latter half of the 1990s, India lived in the knowledge of a nuclear capability, deliverable not only from its eastern but its western side as well. An overt demonstration of India's own nuclear weapons capability in 1998 was a step towards establishing credible nuclear deterrence and preventing the possibility of nuclear coercion or blackmail.

Sandwiched between two adversaries that not only hold individual nuclear arsenals, but also work in close strategic collaboration, if there is any country that needs missile defence, it is India. However, the strengths and limitations of missile defence for India need to be examined in the light of certain factors unique to the Indian situation.

Firstly, given that the missile threats emanating from the neighbourhood are mostly from short or medium or intermediate range missiles, that naturally have shorter trajectories and flight times, the technological parameters of any missile defence that could meet India's requirements would need to be essentially

17. Some Indian strategic analysts such as Air Cmde Jasjit Singh opine that the primary motivation for Beijing's alliance with Islamabad was to counter American sway over the region. In the process, it was also able to use Pakistan to checkmate India. Viewpoint as expressed in a discussion with the author.

different from what the US needs for intercepting ICBMs. The system will have to be technologically far more sophisticated in order to cater for a very fast response to intercept short, medium or intermediate range ballistic missiles. This also has implications for command and control and the requirement for pre-delegation of launch authority, problems that are discussed in the section on operational issues.

Secondly, technological challenges in the case of India are further exacerbated by the fact that one of the adversaries (China) has worked consistently on counter-measures to defeat the American missile defence. China, in fact, has spent most of the last decade working on counter-BMD technologies to defeat the American missile defence. These include electronic jammers, chaff and decoys, use of heat resistant and radar absorbing materials and manoeuvrability of reentry vehicles. While information is scarce on its success rate in these efforts, they will nonetheless stand it in good stead in the case of India, whose BMD would in any case be far more rudimentary than what the US would field. One major challenge for BMD would be correct warhead identification from decoys and debris—a counter-measure easy to incorporate and which can be expected from China. Also, given the Sino-Pak nexus, there is every reason to assume that the benefits of counter-measures will become available to Islamabad, seriously compromising the efficacy of the Indian BMD.[18] In fact, for this reason, the two ABM tests carried out until now can provide little guarantee of their effectiveness in real operations. Test scenarios are always far simpler compared to the myriad parameters that populate operational engagements.

Thirdly, it must also be recognised that a BMD is essentially a

---

18. China has undertaken joint R&D, manufacture and application of small multi-mission satellites in collaboration with Pakistan, Iran and Bangladesh. For more, see Jasjit Singh, "New Space for Arms Race," *Indian Express*, January 27, 2007.

defence *only* against ballistic missiles. Therefore, it is not nuclear defence. It only seeks to provide partial defence against one type of nuclear delivery vehicle. As was said by an American analyst, "A lot of spending on national missile defence right now may make us feel good; but it is rather like putting ten locks on your front door while leaving the back window wide open. It may make some of us feel secure, but fails to account for the fact that there are other (more likely) vehicles for the delivery of nuclear weapons…"[19] In case the adversary decides to use a delivery system other than ballistic missiles to launch a nuclear attack on India, then the BMD would prove ineffective. The accelerated pursuit of nuclear armed cruise missiles by Pakistan (Babur or Hatf 7 with a range of 700 km based on Chinese DH-10 designs, and Ra'ad or Hatf 8 with a range of 350 km and using stealth technology), and their availability with China complicates the scenario by increasing the modes of nuclear delivery. Even in the case of ballistic missiles, use of lofted or depressed trajectories of ballistic missiles would affect BMD effectiveness.

The above is not to suggest that India must not move towards some sort of BMD capability. It only serves to highlight the challenges that an Indian missile defence will have to deal with so that necessary trajectories of technology development can be drawn up. Secondly, it also cautions against placing too much faith and resources in the BMD basket. Of course, there is no denying that some sort of BMD would surely complicate the calculations of the adversary. Just as a demonstrated nuclear capability, however imperfect, provides *existential deterrence*, so the BMD, however lacking in foolproof guarantees, will provide *existential defence*. An adversary planning to launch nuclear missiles against India

19. Bruce MacDonald, "Planning and Options for Missile Defence," in Robert L Pfaltzgraff, Jr., ed., *Security Strategy and Missile Defence* (New Delhi: Ritana Books, 1997), p. 122.

can never be sure that the BMD set up by India would not work, thereby defeating his objective of a decisive, disarming first strike. He would, therefore, always have to account for nuclear retaliation – a basic premise of deterrence.

At a time when China's military capabilities are growing and its intentions remain unclear, laying the groundwork for a military hedge or insurance in the form of BMD that could evolve into a stronger system certainly makes sense. While China is far ahead on counter-measures technologies and should be able to neutralise a nascent Indian BMD, it can yet never be certain that India's BMD will not work. And it is this seed of uncertainty that underpins nuclear deterrence. There are some who argue that Indian pursuit of missile defence would antagonise China and make it respond aggressively by further enhancing its nuclear and missile related developments, as well as increasing its strategic cooperation with Pakistan.[20] However, it must be pointed out that China's enhancement of its nuclear and missile arsenal is more in response to American initiatives and less directed by Indian moves. Therefore, while China will certainly closely monitor nuclear and missile developments in India, it would be relatively less concerned about them as long as its own nuclear deterrence vis-a-vis India remains credible and unaffected by an Indian defence.

At the same time, BMD would impact the Indo-Pak nuclear relationship. In fact, missile defence indicates preparation for deterrence failure, which is more necessary in the case of Pakistan since the Pakistani military perceives its nuclear weapons as providing it with the immunity to blatantly conduct sub-conventional warfare. From an Indian perspective, an effective BMD could

20 For instance, see Harsh V. Pant, "India and Missile Defence: Lull After a Storm," in his book, *Contemporary Debates in Indian and Foreign Security Policy: India Negotiates Its Rise in the International System* (New Delhi: Palgrave Macmillan, 2008), p. 106.

nullify the advantage that Pakistan believes it has acquired with its possession of nuclear weapons and would instill caution in the aggressor. Theoretically, a BMD could untie India's hands and allow some sort of a conventional response to Pakistan-sponsored terrorist activities since Pakistan would be compelled to account for the BMD being able to handle escalation, at least partially. Nevertheless, it must also be said that the BMD cannot guarantee absolute defence against a Pakistani nuclear strike. Therefore, Pakistan will not be actually rendered vulnerable by an Indian BMD because India will still be deterred by the chance of some Pakistani missiles being able to penetrate the shield. As stated by Rajesh Basrur, "The purpose of an Indian missile defence can at best be to try and minimise damage after deterrence has failed, which is far from saying that it will give an Indian leader the confidence to strike first."[21]

Obviously, that is not how Pakistan perceives the development of an Indian BMD. Islamabad decries the move for increasing India's invulnerability and eroding the credibility of Pakistan's nuclear deterrent. In fact, Pakistan's reservations about BMD stem from its larger opposition to the US programme which Gen Musharraf had criticised in 2001 itself as a move that would "jeopardise international stability, trigger a new arms race and undermine international efforts aimed at arms control and disarmament."[22] But, some Pakistani analysts have cautioned their country against being drawn into an arms race with India and have suggested instead that "it would be more appropriate to counter an Indian missile defence with hardened and mobile basing, counter-measures, and a small numerical preponderance in relation to Indian defence capability."[23]

---

21. Rajesh Basrur, "Missile Defence and South Asia: An Indian Perspective" available at http://www.stomson.org/southasia/pdf
22. B. Murlidhara Reddy, "Musharraf Opposes NMD," *The Hindu*, May 13, 2001.
23. Zafar Nawaz Jaspal, "India's Endorsement of the US BMD: Challenges for Regional Stability," *Islamabad Policy Research Institute Journal*, vol.1, no.1,

This is a logical conclusion arrived at by most states grappling with the challenge of missile defence. Accordingly, the responses of Russia, China and Pakistan hinge upon three techniques: plans to saturate the missile defence by expanding own missile force; to defeat it by enhancing own missile force with penetration aids; or to underfly or bypass the defence with cruise missiles, bombers or other non-ballistic missiles. Indeed, the advances in detection and interception capabilities have been sobered by improvements in counter-BMD technologies such as stealth, jamming and camouflage. Of course, these counter-measures—whether a simple increase in the number of missiles or equipping them with greater manoeuvrability or other decoy mechanisms—entail a cost. In fact, the spiral of arms race between offence and defence must not be overlooked. Table 7.1 provides a fair indication of the tussle between counter-measures and counter-counter-measures.

Table 7.1

BMD Counter-Measures	BMD Counter-Counter-Measures
Faster burn times	More interceptors
Jam homing sensors	More missile sensors
Balloon decoys	Better sensors/ More kill vehicles
Multiple launches	More interceptors
Enhanced trajectory manoeuvres	Better tracking
Attack ground stations	Build back-up ground stations
Attack communications	Better communications security

Obviously, a cycle of action and reaction is inevitable and India must step into this field intelligently to avoid being sucked into a potentially counter-productive arms race. At the same time, it might be worth weighing the possibility of using some Indian steps in this direction to deliberately draw Pakistan into an arms race as one way of containing it. The moot point is whether India

Summer 2001, pp.28-43.

can achieve this without burning its own fingures.

Nature of an Indian BMD

Having examined India's strategic interests and rationale for a missile defence as a component of its nuclear strategy, it is necessary that some definite attributes of an Indian BMD must be identified. How elaborate must it be? What should be its mission and, hence, the nature of the associated architecture? Which areas/assets must it defend? What level of system effectiveness could be considered acceptable? And, against what kinds of missiles – ballistic and/or cruise (given the proliferation in latter technologies in the region)? Should it be articulated as a rigid system or can it be made flexible so that it can incorporate system-level responses as technology is indigenously developed or acquired from other sources?

To start with an obvious conclusion, for India, missile defence cannot be possible all over the country, given the sheer technological and financial challenges involved.[24] A cost-benefit analysis makes it difficult to justify such spending on a system that will not be foolproof and will offer a defence only against one kind of a delivery vehicle. Therefore, the missile defence would have to be point or area specific, and, hence, would be required over multiple sites. It would essentially be ground-based defence

24. Prof Dean Wilkening, Director, Science Programme at the Centre for International Science and Arms Control, Stanford University, has done extensive research to establish that defence of all of India is possible through the optimum placement of sophisticated BMD architecture (radars and interceptors). In a presentation of his research findings made at the Centre for Air Power Studies, New Delhi, on Decmber 18, 2008, he estimated that even a *modest* capability to defend against Pakistan would entail a system cost (50 THAAD interceptors, 20 SM-3 Block II interceptors and 2 Green Pine radars) of more than $ 1.5-2 billion. In the case of China, the cost would be upwards of $ 3-4 billion for a system consisting of 50 THAAD interceptors, 100 SM-3 Block II interceptors, 1 PAVE PAWS radar and 1 EMR. In either case, the related command and control infrastructure will also have to be suitably upgraded.

supported by space-based (and some sea-based) capabilities of early warning and communications. Such a capability demands clarity of policy, development, testing, deployment and operation. However, it would be impossible, and even counter-productive, to freeze a final BMD architecture at any point in time. The system would have to evolve in a flexible manner based on technological developments, possibilities of acquisition of certain components/ capabilities from external sources, and inputs received from the developers and the users. What is necessary, nevertheless, is a basic understanding of the assets that BMD must protect, how it must do so and the broad contours of command and control structures necessary for the purpose.

In fact, the first aspect to highlight in a range of issues that engulf BMD is that "interception" may be the most evident and 'glamorous' part of BMD, but it is only one activity in the entire system. It is, in fact, the end product of a long chain of other actions, and only if these are performed well can interception be a success. The BMD system architecture, therefore, has to comprise several systems, as briefly identified below and the technologies supporting each one of these activities is critical for the effective functioning of BMD.

- **Surveillance Systems**: The task of these early warning systems is to detect the launch of a missile soon after the event occurs in order to provide more time for decision-making and launch of interceptors. Advanced detection and tracking requires sophisticated early warning radars placed at critical locations on earth, sea or in space, discrimination algorithms and high power, efficient solid-state transmission and reception modules.[25] Continuous improvement of sensor technologies

25. For more on the technological challenges with respect to command and control, see Daniel McMorrow, "Future Missile Defence Research: Status and Prospects," in Platzgraff, Jr., n.19, p. 73.

in order to improve range, resolution and discrimination capability of surveillance systems is critical.

- **C3 Structure**: Command, control and communications naturally are at the heart of BMD. This is required to be based on technologies that allow quick interpretation and analysis of information received from sensors and an effective and redundant communication architecture to handle rapid flow of information to and from the decision-maker. Automated decision aids, data fusion, high data rate communications are essential prerequisites for this activity. Equally important, however, is clarity on chain of command and execution of decision—who would have the weapon release authority and how it would be communicated to the man on the ground.
- **Interceptors**: This comprises the actual vehicle that would target the incoming missile and hit it before the enemy missile reaches its own target. This naturally requires the interceptor to be faster, more manoeuvrable and accurate than the target missile, and capable of discriminating between chaff and decoys a major technological hurdle.

As is evident from the above, BMD technology necessitates the use of space for several support functions. From surveillance through satellite-based assets, and from launch of interceptor to navigation through GPS are some activities that can be best conducted from space. It is for this reason that the US envisages the use of this medium for its BMD programme, including at some time in the future for the placement of interceptors so that they could undertake global missions. Not surprisingly, therefore, there is linkage between BMD and the eventual weaponisation of space, once the technologies are adequately mature and the politics has so evolved to accept such an eventuality. It is imperative that India take adequate notice of these linkages and draw up its own responses in terms of technology developments and negotiating strategies in relevant international fora deliberating on these issues. While weaponisation of space is certainly not in India's interest, the enhanced use of space for national security in general, and BMD in particular, must be exploited.

Operational Issues

Where?

The missile defence debate in the US was driven by the president's stated desire to defend the entire continental US, deployed forces, and friends and allies. The idea was to establish an astrodome type shield that could provide an umbrella cover to the entire nation and its various extensions. The goals have since been tempered to more practical limits. In the case of India, fortunately, there are no such inflated expectations from BMD. While it is logical for the politicians to express their desire to maximise defence of the populace, it is clear that it will not be practical, nor affordable, to provide missile defence over all population centres, even if technologically feasible. In fact, since deploying MD is not a cheap option, nor will it be extensively available for deployment in the short term (considering industrial production limits), prioritisation on the basis of strict criteria will have to be undertaken. Logically, among the counter-value targets, the national capital, and major financial and industrial hubs should qualify for some sort of defence against nuclear capable missiles.

At the same time, India's no first use nuclear doctrine, compels reliance on survivability of nuclear assets vital for retaliation. Missile defence over some strategic sites and assets such as the Nuclear Command Post and the Alternate NCP, as well as air bases and ports from where nuclear capable platforms would operate, key military forces, nuclear installations (despite the existing Indo-Pakistan agreement not to attack each other's nuclear facilities), critical infrastructure and industrial units is a possibility. As is evident, this list is only illustrative.

How?

The first phase of BMD deployment in India is planned to be ground-based. The interceptor missiles will be dispersed at strategic locations decided on the basis of likely direction and trajectory of incoming enemy missiles. Sensors would also be largely on the ground, but the use of space for surveillance and communications would certainly expand as India's military use of the medium increases. The possibility of sea-based missile defence is worth exploring. It certainly provides the obvious advantage of enabling forward location of sensors and shooters closer to launch sites for early detection and interception. But in the case of Pakistan, and especially China, missile launch sites are far away from the seas. Also, sea-based missile defence suffers from some limitations. For one, the ships operating in a forward location are far more vulnerable to enemy attack themselves and their own protection requires adequate defence, including through the presence of additional escort vessels in the region. This necessitates adequate numbers of ships for effective forward deployments. Also, it has been pointed out that "conducting BMD operations may reduce a ship's ability to conduct air-defence operations against aircraft and cruise missiles due to limits on ship radar abilities. BMD interceptors occupy ship weapon-launch tubes that might otherwise be used for air-defence, land-attack, or anti-submarine weapons."[26]

---

26. Ronald O'Rourke, "Sea-Based BMD – Background and Issues for Congress," *Congressional Research Service Report for Congress,* December 4, 2006.

By Whom?

While the DRDO is the basic building block of BMD and responsible for its development, testing and evaluation, at the operational level, however, given that air defence of the nation is essentially an air force function, some opine that missile defence should also logically be entrusted to the same Service. This also appears rational given that the Indian Air Force already runs an elaborate network of early warning systems and other sensors, including AWACS as and when it will be integrated, that will facilitate early detection of incoming enemy missiles. However, the air force would have to perform this task jointly with the other sister Services in order to avoid duplication of capabilities and to optimally use on hand systems along with new installations without disrupting existing operational patterns of command and control. Missile defence would require the surveillance and the elaborate, secure and redundant communication systems that all the three Services have developed. As has been explained by Gp Capt A.K. Singh:[27]

> The Army has a robust communication network in the North and Northeast, and the Navy has its own surveillance and communication resources along the coastline and over seas, while the IAF has the largest radar and communication network across the nation and island territories, and all these resources would have to be integrated into an effective early warning and interception network for BMD.

Having identified the major players in the exercise of BMD, it is necessary to establish the authority that could sanction release of the interceptor. At one level, unlike the nuclear weapons that are strategic tools of political utility whose use must be sanctioned

27. Gp Capt A.K. Singh, *Transformation of Air Defence in Asia* (New Delhi: Knowledge World, 2008), pp.170-171.

by the political authority, BMD is a weapon system for military use in defence. To that extent, its command and control must rest with operational military commanders, especially also because the timeline for decisions on launch is so short. In any case, the interceptors will not be armed with nuclear warheads. However, since the BMD is closely related to nuclear strategy, and affects, as is also affected by decisions, on offensive strikes, its use would have wider implications and, hence, the political leadership might require to be convinced of the need for pre-delegation of authority. In fact, it would be necessary to engage with the Political Council on this as a matter of peace-time planning in order to receive approved guidance for near automatic execution in case of a crisis.

Conclusion

As of now, it is claimed by the DRDO that India would have a BMD capability within three years. In the first phase that would be completed by 2010-11, the system would be able to tackle 25 incoming missiles of 2,000 km range simultaneously through the firing of two missiles per incoming missile to intercept the target at 45-50 km and 15-20 km altitudes in the space. In the next phase, it would also develop the capability to intercept missiles up to 5,000 km range as also cruise missiles.[28] However, as has been stated earlier in the chapter, India must remain realistic in its ambition of BMD and the ability of the system to handle the threats specific to the country. The requisite amount of testing of all components and sub-systems of the system to ensure robust performance in real engagement environments and against simple counter-measures is necessary. This must continue at a steady pace and without too

---

28. "India to Build ABM Shield by 2011", http://www.kalingatimes.com, January 25, 2008. Also see Jangveer Singh, "BMD Shield for India in Three Years," *Tribune*, January 27, 2008.

much fanfare so as not to oversell capabilities and raise unrealistic expectations at home, nor destabilise the region.

Another aspect that must be given adequate thought is that of external assistance for the BMD programme. For some time now, the US has been keen on some sort of a partnership with India on BMD. This was one of the four issues taken up as part of the Next Steps in Strategic Partnership (NSSP) and more recently, the two governments have exchanged briefings and viewpoints on the subject as part of the Joint Technical Group, a sub-group of the overall Indo-US Defence Policy Group. Several American companies such as Boeing, Lockheed, Raytheon and Northrop would certainly be eager to seek customers for their products as well as partners for some of their ongoing or planned projects. With the problems facing the American economy at the moment and the likelihood of strict governmental cuts on future R&D, it is not unfeasible that India and its relatively advanced scientific and technological developments will prove to be attractive to US defence companies. While there can be no denying that incorporating imported elements such as mobile radars, sensors or other relevant technologies could bolster India's indigenous capabilities, these steps must be taken after serious consideration and after making optimum use of the leverages available with a resurgent India today. Dr Saraswat is on record to have claimed that the BMD system under development in India, and to be operated in tandem with the medium range Akash surface-to-air missile (SAM), is superior to the US Patriot Advanced Capability (PAC)-3 in terms of range or altitude of interception.[29] Such confidence has compelled American companies to take note of the Indian advancements and they have offered to "blend in" their

29 "Indo-US Defence Ties: Tasting the Lockheed Blend", www.domain-b.com/industry/defence, February 8, 2008, as reproduced in *Nuclear Security*, vol. 2,no. 8, February 15, 2008.

technologies with Indian systems rather than offer outright sale of the system. It might, therefore, make sense to collaborate with some of the more experienced companies in the field of BMD infrastructure such as C3 components and radars. In any case, India has been developing some of these systems with Israeli and French help. The Israeli Green Pine early warning radars have been upgraded to meet India's purpose of the Long Range Tracking Radars (LRTR), which according to Dr Saraswat is "30-40% better in target classification and identification. It can prioritise the incoming threats and track 200 targets simultaneously."[30]

Given the rapid Indian indigenous advancements in this field, it is critical that decisions on foreign collaborations or acquisitions be made after giving adequate thought to cost and political considerations. The purpose of BMD for India, its advantages and limitations must be adequately understood before rushing into hasty decisions. Also, as the DRDO moves towards providing a deployable and operational BMD for the nation, it is imperative that greater thought be devoted to operational issues, especially by the Services. The immediate objective must be to ensure that as missile defence capabilities evolve, senior political decision-makers and military commanders become familiar with the systems and the operational challenges and strategic issues involved. The lure of the shield demands utmost rigour in thought and action.

30. Rajat Pandit, "India on Way to Joining Exclusive BMD Club," *The Times of India,* November 26, 2007.

8

Material Matters

Credible Nuclear Deterrence and Fissile Material Concerns

As a state with nuclear weapons, fissile material is central to India's nuclear strategy for the simple reason that it is the basic raw material for creating nuclear weapons. Engaged as the country is in the process of building a credible nuclear deterrent, it has to ensure that the requisite amount of fissile material is available to create an arsenal that has been determined as necessary by the decision-makers. India has refrained from identifying what that number might be, except to qualify it as "minimum". However, whatever that unidentified quantity may be, it certainly would require a commensurate amount of fissile material. This implies two things: one, India *must ensure the availability of the required fissile material* through the necessary technological processes such as reprocessing or enrichment, and through the necessary diplomatic positions so that no non-proliferation/arms control obligations put a stop to the country's ability to stockpile fissile material before it has accumulated what it deems essential; and, second, India *need not accumulate an unnecessarily large quantity of the material* in excess of its pre-determined requirement for credible deterrence. While there are several who argue that given the mercurial nature of threat perceptions, no government can foreclose decisions for the future, it may, however, be said that a well thought out threat assessment could allow the government of the moment to take a considered decision on the matter.

It is, therefore, imperative to examine and understand the inter-linkages between India's fissile material stockpile, the likelihood of the conclusion and the possible contours of a Fissile Material Cut-off Treaty (FMCT), and the requirements of India's credible nuclear deterrence. For the time being, an FMCT, in a ready to subscribe format, appears a fair distance away. In fact, the flurry of activity created by the USA in May 2006 with its sudden and rather unexpected presentation of a draft of the treaty in the Conference on Disarmament (CD), during the conduct of week-long thematic debates in the CD's second session of work in April-May 2006, seems to have died down. The American initiative could only briefly animate the 65-member CD, which has been experiencing a decade-long period of lull during which the body has not even been able to agree on a programme of work.

Several reasons were attributed for the US action – a possible display of faith in multilateralism[1] in the wake of criticism of its unilateral initiatives in Iraq; an attempt to prove its disarmament credentials in view of the forthcoming NPT Review Conference in 2010; a move to quell some of the criticism on its development and deployment of missile defence and refinement of its nuclear weapons infrastructure; or simply an attempt to revive the defunct

1. It is noteworthy that in his statement in the CD before presenting the US draft, Stephen Rademaker, Acting Assistant Secretary, Bureau of International Security and Non-proliferation emphasised US conviction in "effective multilateralism," For text, see CD website http://www.unog.ch

CD, or at least deflect blame in the case of its possible demise owing to its long time inaction.

Whatever the reasons, the issue of proscribing fissile material production did appear to make a comeback on the international negotiating table, albeit only briefly. At the time of writing this chapter, it appears that the journey of an FMCT will prove to be long and tortuous in the CD. Of course, post-2006, a special coordinator has been appointed to initiate discussions. But little substantive movement has been witnessed since then and countries have committed to no more than studying/examining the various proposals before them. Therefore, in the CD, the fate of an FMCT appears uncertain. However, there is a possibility, not particularly bright but extant nevertheless, that in case there is consensus within the P-5 on the necessity of an FMCT, then they might conclude one within themselves and then put it out for signatures of the other states with nuclear weapons (basically, the three NPT outliers, and North Korea whose status in the NPT remains ambiguous). Given the difficulties in consensus building within the CD, the P-5 might find this a quicker and more effective route to conclude the FMCT, provided they are themselves convinced about the need for such a treaty. In such case, it may emerge more quickly than anticipated.

In view of such developments that could suddenly confront the country, it is necessary that India must closely monitor the developments in the CD and outside, participate in the process where it can, and continuously assess what kind of an FMCT would suit India best. It is certain that at some point in the future, and possibly within the term of the new Democrat Administration led by Barack Obama, the debate over the FMCT would revive. India will then have to take a position and it may be recalled that under the terms of the Indo-US Joint Statement 2005, and the civilian nuclear cooperation agreement signed thereafter, India is committed to working with the US for the early conclusion of an FMCT.

Therefore, some questions must be considered and answered now. For instance, what would be the strategic implications of a stop on future fissile material production on India's nuclear deterrence? How would the FMCT impact India's unique nuclear fuel cycle activities? Is India in agreement with the broad definitional concepts of what constitutes fissile material? Is verification under the FMCT in India's interest or not? The range of questions is diverse, encompassing both political and technical evaluations. This chapter considers four specific issues related to the FMCT: definitions, scope, verification and implications of each of these on India's strategic capability. It seeks to draw the connection between fissile material concerns and India's peculiar brand of credible minimum deterrence in order to explain why India's fissile material needs must be finite and that there is no need for a runaway stockpile accumulation.

Definitions

According to the most commonly accepted definition of "fissile material", it includes those materials that are fissionable when irradiated by both slow or thermal neutrons.[2] The IAEA Statute defines special fissionable materials as plutonium-239 (Pu-239), uranium-233 (U-233), uranium enriched in the isotopes 235 or 233 which means uranium containing either or both of these isotopes in an amount such that the abundance ratio of the sum of these isotopes to the isotope 238 is greater than the ratio of the isotope 235 to the isotope 238 occurring in nature. Staying with this interpretation, the US draft treaty of 2006 defines fissile material as "Pu except Pu whose isotopic composition includes 80

2. Brian G. Chow, Kenneth A. Solomon, "Limiting the Spread of Weapon-Usable Fissile Materials," *RAND Report*, 1993, p.1.

per cent or greater Pu-238; uranium containing a 20 per cent or greater enrichment in the isotopes U-233 or U-235, separately or in combination."

It is important to have clarity on these definitions since the scope of verification, if the FMCT emerges as a verifiable treaty, shall be determined on this basis. While logically speaking, all facilities and activities that could be used to produce fissile material (such as military designated plants, civilian power reactors that produce Pu or U-233) or divert weapons usable fissile material (such as tritium producing reactors that use HEU as fuel that could be diverted for weapons) should be subject to FMCT proscription or verification. However, ambiguities arise due to the dual use nature of such activities.

One crucial question in this context is whether the definition of fissile material should be restricted to only "weapon-grade" materials such as HEU and Pu-239, or whether it should also include "weapon usable" materials such as tritium as well. Technically, tritium is not a fissile material, but is used in fission weapons to enhance their yield and in thermonuclear devices as a fission trigger.[3] Logically speaking, if the FMCT is truly to be a step towards capping nuclear arms production, then it must include tritium as well. But the US has objected to its inclusion on the grounds that bringing in tritium under the fissban would make the treaty a "weapons ban", not a fissile material cut-off. Tritium decays at the rate of about 5 per cent per year and, hence, all the NWS need it periodically to maintain operational readiness of these weapons.

What does tritium mean for the Indian nuclear arsenal? Having demonstrated the capability of conducting boosted fission and thermonuclear explosions in May 1998, India would need tritium

3 T.S. Gopi Rethinraj, *Jane's Intelligence Review*, January 1998, describes it as a "strategic material in the creation of thermonuclear weaponry."

in both devices. In fact, as long as India is required to maintain an effective, credible and reliable nuclear deterrent, tritium would be necessary. On this issue, therefore, India should go along with the other NWS.

The second contentious definitional issue relates to describing the exact activities that should be designated as engaged in "production of fissile material." This would determine whether only reprocessing and uranium enrichment facilities – the two primary routes to fissile material production—should be defined thus, or whether other facilities that could be potential sources of fissile material, such as nuclear power reactors and spent fuel reprocessing plants too be treated as fissile material production facilities.

For India, this definition has particular implications given its unique conceptualisation of a three-stage nuclear programme for electricity generation. In the first stage, PHWRs produce plutonium, which after reprocessing is used as fuel in FBRs. These, in turn, produce more of the same and U-233 that would be used with thorium in the third stage of the programme. Therefore, fissile material is involved in every stage, either as fuel or product, or both. As of now, India has no plan of placing its first prototype FBR being constructed in Kalpakkam and likely to become operational by 2010 under safeguards. However, once an FMCT enters into force and India subscribes to it, it will be obliged to accept safeguards on its FBRs.

In fact, two issues need to be considered in this context. Firstly, whether an FMCT *would allow plutonium production for civilian purposes*? This certainly appears most likely since several countries, inluding France, Russia, Japan, and Germany support commercial reprocessing of plutonium. In fact, even the USA that had adopted the once through fuel cycle since the days of the Carter Administration, expressed greater interest in reprocessing during the Bush Jr. Administration. So, the FMCT would certainly allow the production of plutonium for civilian purposes.

The second issue pertains to the assurance that the *plutonium so produced is not diverted for proscribed military use*. Ideally, a safeguards regime based on proper material accounting and verification should provide this guarantee. In order to be effective, such a regime would call for immense resources – financial, manpower, technical and political. But, as is discussed in the section on verification, there is no consensus on inclusion of provisions for verification in the FMCT. In such case, every nation would only be required to make the necessary declarations on fissile material being produced in their country for peaceful purposes. If reprocessing for nuclear power production is accepted as a legitimate activity, then one would equally have to be prepared to accept the risks arising from the proliferation potential of this activity since reprocessing can be undertaken in facilities that are simple and small and, hence, could escape detection. A more considered evaluation of India's position on verification is undertaken elsewhere in the chapter.

Scope

Conflicting viewpoints over the scope of the treaty have long existed. There is no consensus over whether it should be confined only to future production of fissile material, or include existing stockpiles of fissile material also within its purview.[4]

The five NWS (as also India) favour a 'look ahead' orientation

4. The divergent opinions are naturally also reflected in how countries refer to the treaty. While India continues to use the acronym FMCT since it argues that the treaty must only prohibit future production, other countries have used different terminologies as per their country positions. Interestingly, Pakistan, that argues for the inclusion of existing stockpiles, prefers to use "FMT" and leaves out the "cut-off". Several others also use the more general term "fissban" so as not to prejudge the issue of scope and stocks.

by limiting the FMCT to verifying only a future cut-off in fissile material production. However, countries such as Egypt, Pakistan, Iran and Algeria argue in favour of treaty provisions that would enable reductions in existing stockpiles. On June 9, 1994, the Pakistani representative at the CD proposed a schedule for the progressive transfer of existing stocks to safeguards, (transfers first being made by states with the largest stockpiles) so that unsafeguarded stocks could be "equalised" at the lowest possible level.[5]

In fact, the negotiations found themselves deadlocked over this issue in the mid-1990s, leading Ambassador Shannon to concede in his report of March 24, 1995, that the "mandate for the establishment of the ad hoc committee does not preclude any delegation from raising for consideration in the ad hoc committee any of the above mentioned issues."[6] Hence, he tried to paper over the differences to put negotiations back on track by not excluding any viewpoint.

The US draft of 2006, however, clearly puts forth a "straightforward scope" that will ban production of fissile material after the treaty has come into force. It accords no retrospective value and does not include existing stockpiles. This is exactly what India desires for the obvious reason that it would impinge on the size of the nuclear arsenal that India could build with the accumulated fissile material that has not been turned into weapon cores at the time of the entry into force of the FMCT. Pakistan, on the other hand, for the reason of not "freezing" nuclear asymmetry, and

5. CD/PV/681, June 9, 1994. For more on recent Pakistani position, see Gaurav Rajen and Michael Vannoni, "Fissile Materials Control in South Asia: Regional Analyses and Potential Confidence Building Measures," Cooperative Monitoring Center Occasional Paper, SAND 2006-0432 (Albuquerque: Sandia National Labs, February 2006). Pakistan's Ambassador to the CD, Munir Akram, had reiterated in 1998, "We cannot agree to freeze inequality," August 11, 1998.

6. Report of Ambassador Shannon to the CD, March 23, 1995.

more likely, for buying time, and to showcase its solidarity with the Islamic countries, continues to maintain that the FMCT must consider "past, present and future" production of fissile material. However, given that the P-5 do not share this view, the FMCT is unlikely to bring in the existing stocks under its purview.

Secondly, as per the 2006 US draft, fissile material production for non-explosive purposes such as naval propulsion is not included in its scope. This too is in India's interest. Given its retaliation only counter-strike nuclear doctrine, India will need to build its credible nuclear deterrence on a robust retaliatory capability to be provided by nuclear-powered submarines capable of carrying nuclear tipped missiles.

Verification

As a matter of tradition, and habit, arms control measures from the Cold War era have always been anchored in some sort of verification mechanisms. The first major treaty to break this trend was the Biological and Toxic Weapons Convention (BTWC) that has no verification arrangements. Would the FMCT follow the BTWC model or be made verifiable like the Chemical Weapons Convention (CWC) is still an open question. The complexity of definitional parameters, conflicts over scope, and sensitivities of state sovereignty have led to divergent viewpoints on the need and feasibility of verification.

During the limited negotiations held so far, many options have been proposed ranging from "focussed" to "wide-scope" to "no" verification. The first envisages safeguards only on sensitive fissile material production facilities such as enrichment and reprocessing plants, on fissile material produced after the treaty's entry into force, and on facilities where these materials are present. Wide-scope verification, meanwhile, entails the inclusion of even less sensitive civilian facilities such as fuel fabrication plants, power reactors and declared shutdown production facilities, declared operating production facilities and also a look-out for undeclared facilities through environmental monitoring, challenge inspections and remote sensing.

Besides the scope of verification and its technical feasibility, its *political acceptance*, especially by NWS, is no less a challenging proposition. Problems would arise in situations where sensitive production facilities to be safeguarded under FMCT happen to be co-located with those where military activities such as the assembly of nuclear weapons could be in progress, or where undeclared stockpiles from past production were being held. The verification system would also have to be adequately supported by the establishment of national material accounting in order to determine the accuracy and veracity of the accounting system. Moreover, in case uniform verification arrangements were not prescribed for all states, then the *de facto* nuclear, but *de jure* non-nuclear status of India, Pakistan and Israel would create further complications.

It was in view of the complexities and practical problems related to verification of the FMCT, that the US had stated in the UN First Committee Plenary Session on October 16, 2004, that it found it fruitless to negotiate verification procedures since it would only cause unnecessary delays.[7] Ambassador Jackie Sanders, head

7. Though the Bush Administration did not explain the specific concerns that led to the conclusions of its inter-agency review, these included (a) the fear that far-reaching inspection rights would be needed to uncover possible covert nuclear material production facilities, and that giving other countries such inspection rights in the US might pose a danger of compromising US secrets; (b) even such a far-reaching inspection regime might not be able to provide high confidence that covert facilities would be detected before they could be producing nuclear bomb material; and (c) such a wide-ranging inspection regime would be expensive. The Administration had offered much the same reasons for ending negotiation of a monitoring protocol for the BTWC.

of the US delegation, had said, "The objective of an FMCT is not its verification, but the creation of an observed norm against the production of fissile material intended for weapons."[8] She emphasised that since the FMCT would ban production of fissile material for one type of activity (use in nuclear weapons), while allowing production for other activities, even when undeclared fissile material was found in a state under safeguards, it would be difficult to establish non-compliance, unless the date and purpose of production could be conclusively established and this was not going to be easy or even possible. Whether wide-scope or focussed, in either case, FMCT verification "would require an inspection regime so extensive that it could compromise key signatories' core national security interests and so costly that many countries will be hesitant to accept it."[9] And yet, the regime may still not be able to grant high confidence of compliance.

Hence, the US found it prudent to do away with safeguards *in toto*, in favour of using national technical means (NTM) for detecting non-compliance. Accordingly, its draft text vests the primary responsibility for verification with individual nations using their NTM, and equipped with the right to bring violations to the notice of the other treaty members as also of the UN Security Council.

None of the other major players in the FMCT debate has expressed much support for this option. China has vaguely stated, "It is necessary to have the verification issue carefully explored

---

8. http://www.state.gov/t/ac/rls/rm/2004/37639.htm Released on October 21, 2004

9. Even in the year 1995, preliminary cost estimates of verification had calculated a doubling or tripling of IAEA's annual safeguards budget, then at about $ 70 million. It was also expected to require 7-11 per cent annual growth over 10 years in trained inspectors and person days of inspection effort. For more, see Steve Fetter and Frank von Hippel, "A Step by Step Approach to a Global Fissile Material Cut-off," *Arms Control Today*, vol.25, no.8, October 1995, p.5.

and prudently handled. Opinions of all parties concerned, including the specific considerations of those who advocate an FMCT without verification must be taken on board." It is no secret that China is averse to intrusive inspections of any kind. It will have reservations on inspections of its erstwhile military nuclear facilities, especially if they happen to be co-located with civilian facilities requiring verification. Therefore, Beijing would prefer less intrusive safeguards and could respond positively to the US proposal of no verification. However, as always, seeking to keep all options open for a necessary trade-off in the future, Beijing maintains an ambiguous position on the subject.

Pakistan, meanwhile, like India, upholds the need for a verifiable FMCT. Speaking at the CD, Pakistan's Ambassador emphatically stated that a fissile material treaty "*sans* verification *sans* stocks will be *sans* treaty."[10] He opined that a verifiable FMT[11] would be able to: control the spread of nuclear materials, enhance the proportion of weapon usable material under international safeguards and strengthen nuclear export control and reduce the discrimination in the present NPT regime. He also highlighted that better monitoring and accountability of fissile material stocks would block terrorists' access to fissile material.

India too has traditionally supported verification. As Ambassador Jayant Prasad stated at the CD in May-June 2006, "Absence of a verification mechanism may engender lack of confidence in compliance with the treaty, encourage willful non-compliance, and lead to allegations and counter allegations of

---

10. As quoted in Sharon Squassoni, "Banning Fissile Material Production for Nuclear Weapons: Prospects for a Treaty (FMCT)," *CRS Report for Congress*, July 14, 2006.
11. In consonance with its view of inclusion of existing stocks as part of the treaty, Pakistan prefers to call it a Fissile Material Treaty instead of a Fissile Material Cut-off Treaty.

noncompliance."[12] In the absence of its own NTM capabilities, India is uncomfortable with dependence on the US NTM for verification. It is feared that this would boil down to the US deciding who is in compliance and who is cheating.[13] This would not only reinforce the US role as globocop and leave other states vulnerable to American accusations, but also force American determination of non-compliance on the rest of the world. And going by past experience, this has been applied in a selective manner and with less than foolproof intelligence. For instance, while the US non-proliferation radar managed to pick up signals of proliferation in Iraq, and now Iran, it never registered a blip in the case of the Sino-Pak nexus or even in the face of alarming revelations about the scope, reach and expanse of the A.Q. Khan network operating from Pakistan.

However, instead of rigidly sticking to a position only because India has supported it in the past, it would be worthwhile to objectively examine the efficacy of verification under the FMCT. As far as India is concerned, two aspects of verification must be examined: how much confidence would it provide that China and Pakistan are not proliferating, individually, or one to another; and secondly, implications for India's own strategic and civil nuclear programme. As regards the first issue, would a *verifiable* treaty significantly influence China or Pakistan's behaviour? Given their track record, and the primacy of political considerations over proliferation concerns as exhibited by the US and others, can India afford the luxury of being complacent once a verifiable FMCT is in place? Co-location of military and civilian facilities, inability to determine the date of production of fissile material, or its purpose, are only some of the technical loopholes that could be exploited

12. http://www.mea.gov.in, May-June 2006

13. K. Subrahmanyam, "Nuclear Wake-up Call," *The Times of India,* May 29, 2006.

even with safeguards in place. Political considerations would further determine whether even an 'independent' monitoring body would be allowed to level and investigate accusations against China, a rising power, and Pakistan, a strategic ally of the US. Would a verification mechanism then be able to perform the "dual purpose of detection and deterrence" that India seeks?

Meanwhile, what would verification mean for India's own nuclear programme? Irrespective of whether these are "focussed" or "wide-scope", it would mean India having to open up a largely unsafeguarded programme to international inspections. In the case of the Indo-US nuclear deal, India was able to determine its own separation plan and dates of implementation. Such a luxury will not be accorded by the FMCT, placing pressure on India to open up facilities producing or using fissile material. A proper system of material accounting would also need to be established, leaving little scope for plutonium produced in FBRs, presently out of the safeguards regime, being used for weapons production. At the same time, all states with nuclear weapons, including India, would most certainly have to provide the resources, money and material, to share the burden for creation of necessary verification infrastructure.

In short, while truly effective verification mechanisms are indeed laudable and welcome, these would be technically and politically daunting to create. With the kind of safeguards that will be possible, India would have no guarantee that its neighbours are not indulging in proscribed activities. After all, violations of the CWC, despite being verifiable, are not unknown. At the same time, India would have to accept verifications on all its own nuclear facilities. Therefore, India can afford to show greater flexibility on the issue of verification. In case it seeks to delay negotiations and buy time for further fissile material accumulation, insistence on verifications would be useful. But in case major countries appear to be coalescing around a treaty without verifications, then India can let go of its insistence on verification and even earn the goodwill of not standing in the way of the treaty. Meanwhile, at its own end, it is critical that India steadily improves its own intelligence and other NTM capabilities. These would more than make up for a non-verifiable FMCT and also support other bilateral arms control in the region as and when any meaningful measures are concluded between India-Pakistan or India-China.

Indian Brand of Nuclear Deterrence

The reason why an FMCT would have an implication for India's nuclear deterrence is the simple fact that once it comes into force, it would cap any quantitative increase in India's nuclear weapons count. Hence, the urgent need to ensure that the country has enough fissile material for the kind of arsenal that it will need for its stated nuclear doctrine in the short and long terms. This assessment must be based on a few factors:

(a) the role that nuclear weapons are designated in the country's security strategy. In the case of India, this has been clearly spelt out as being political to credibly deter the adversary's use or threat of use of its nuclear weapons for blackmail or coercion. A military role for nuclear war-fighting is ruled out;

(b) the likely source, scope and probability of *nuclear threats* to India's security and the availability of requisite arsenal to credibly deter them. This implies the need for a convincing counter-strike capability, with assured retaliation to cause unacceptable damage to the adversary;

(c) the state of other factors that impinge on the credibility of nuclear deterrence, such as:

  (i) reliability of the warhead design so that it explodes when required and provides the promised yield;

(ii) range, reliability and accuracy of the delivery system so that it hits the pre-determined target;

(iii) robustness of C3I systems, with built-in redundancy for a counter-strike strategy;

(iv) survivability of the nuclear arsenal in its entirety;

(v) operational efficiency of the total nuclear arsenal and its sub-systems, including the human component;

(vi) political resolve for retaliation.

A calculation of fissile material/warheads needed for credible deterrence will require viable data on each of the above aspects. Their cumulative result can then be the basis for rigorous systems analyses and empirical studies for a correct estimate of requirements. This would, of course, have to be a classified exercise, best coducted by those who are in the know. However, this chapter takes a more analytical and less empirical approach to the fissile material requirements for credible nuclear deterrence, and considers some of the relevant issues in the following paragraphs.

The primary objective of India's nuclear weapons as lucidly stated in the country's nuclear doctrine is that of nuclear deterrence. Unlike some other nations that envisage a war-fighting role for their nuclear arsenals, India perceives it only in the political role of deterrence, simply because the weapon is so unique by virtue of its enormous destructive potential that it defies logic to plan to attain any rational political end through its use. Therefore, the nuclear weapon exists to insure the country against blackmail, threat or coercion, rather than for any offensive/aggressive designs. Deterrence, then, is the name of the game for India. The credibility of this deterrence rests not on a large number of nuclear weapons but a realistic minimum number that can effectively deter an adversary from using its own nuclear weapons because he would have to contend with certain and punishing retaliation. In fact, as was explained by Jaswant Singh, India's External Affairs Minister at the time of the nuclear tests,

> In our view, the principal role of nuclear weapon is to deter their use by an adversary. For this, India needs only that strategic minimum which is credible... therefore, the question of an arsenal larger than that of country X or Y becomes a nonquestion. For India, the question is only one of *adequacy that is credible* and thus defines our 'minimum' .[14]

In keeping with this philosophy, the national security establishment has understandably refrained from defining a specific number as minimum since this determination has to be a function of the threat perception, technological abilities of defence and interception, doctrinal pronouncements, damage tolerance threshold assessments and cultural propensities of the adversary at a given moment. As one or more of these parameters change, so could the requirements of India's nuclear forces, though they are unlikely to swing significantly far from a couple of hundreds at the most.

For the sake of analysis, it may be assumed that unacceptable damage for Pakistan would constitute no more than four or five 20 kt weapons each on 5-6 major cities and roughly the same number on about 10 Chinese cities/strategic targets. The damage wrought on modern mega cities, with their high densities of population and their increasingly glass buildings, by a ground burst of a 20 kt weapon would render fatalities and casualties in millions – many orders of magnitude more than Hiroshima or Nagasaki. It would be well beyond the capacities of these countries, with their less than adequate medical infrastructure, to look after the survivors. The

---

14. "India Not to Engage in a N-Arms Race: Jaswant", cited in Ashley Tellis, *India's Emerging Nuclear Posture: Between Recessed Deterrence and Ready Arsenal* (Santa Monica: RAND, 2001). Emphasis added.

material damage would make areas uninhabitable for decades and set the countries back in their economic development by many years. In fact, progressive economic development and betterment of infrastructure and quality of life would lower damage tolerance levels, thereby reducing India's deterrence needs. At the same time, the adversary's future surveillance, reconnaissance and intelligence gathering, as well as precision strike capabilities would be far better and able to inflict a more effective first strike on India, thereby increasing India's deterrence needs. However, the moot point is that these two factors do not necessarily balance out the number of nuclear warheads required. While in the first case, less may be enough, in the second case, more may not be better *unless* the nuclear force (warheads, delivery vehicles and C2) is made more *survivable to mount assured retaliation*. And this emphasis on making the nuclear force survivable will have to be the *mantra* for India in the coming decades.

Yet, for the sake of a hypothetical calculation, even if the earlier mentioned numbers are taken as a rough estimate, then the real number of nuclear weapons available with India *for accurate delivery after the attacker's first strike* has been absorbed should be at least 40-50. If we accept a theoretical/notional figure of at least 25 per cent of the nuclear force surviving a first strike by a second tier nuclear weapon state (China), and a little more in the case of a less capable nuclear state (Pakistan), then for a requirement of at least 50 warheads surviving, we would need a minimum of 200 warheads and requisite number of delivery systems to begin with. An important parameter impinging upon this calculation would be the adversary's ability to degrade the Indian arsenal through neutralising/destroying its C3I systems (through ASAT, cyber attack, kinetic strikes, etc), delivery and launch systems and nuclear warheads, as also his ability to carry out a quick strike damage assessment in order to calculate the residual capabilities of the victim to launch a second strike before the counter-strike can be mounted. Yet another parameter for calculating the number would be the expected reliability of the warhead and delivery vehicle – to reach the intended target and detonate with the estimated yield. A certain allowance for malfunctioning would have to be made.

Credible Deterrence – More than Mere Numbers

For India, credible nuclear deterrence will necessarily have to be a function of two factors: a realistic assessment of what would constitute retributive *unacceptable punishment* for the adversary; and the *level of survivability* that can be ensured for India's own arsenal after absorbing a first strike. The number of nuclear weapons that India must have for credible nuclear deterrence will have to be determined on the basis of both these parameters. An unusually large number in the absence of both these determinations would be useless, just as not enough to meet these two parameters would make deterrence weak and non-credible.

The question that then arises is whether India has stockpiled enough fissile material to meet these two parameters. Given the high level of opacity, ambiguity and secrecy that is maintained on these issues (which itself contributes to deterrence), it is impossible to answer this question with any amount of certainty without specific accurate data. Nevertheless, Western and Indian scientists and strategic analysts have made calculated guesstimates on the basis of which some intelligent conjectures are possible.

For India, Pu-239 is the most significant fissile material available for weapons production. This is currently produced in the two research reactors: the 40 MWth CIRUS[15] and the 100 MWth

15. CIRUS became operational in 1963. It went in for a major refurbishment from October 1997 onwards and resumed operation only in October 2003.

Dhruva[16]. It may also be obtained from PHWRs in two ways: either by reprocessing of spent fuel of PHWRs; or by running the PHWRs at low burn-ups at about 2000 MWd/t instead of the usual 7,000-8,000 MWd/t. This would, of course, reduce their power output, rendering them uneconomical from the point of view of electricity generation, but nevertheless, it would make them available for strategic use.

India does have a small uranium enrichment programme that began in the 1980s. A pilot scale enrichment plant[17] at the Bhabha Atomic Research Centre (BARC) is primarily engaged in research and development activities. A larger production scale facility at Rattenhalli[18], near Mysore, produces not very significant quantities of this fissile material. In fact, a RAND study of the year 2000 suggested that India could produce only 10 kg of HEU per year.[19] A more recent estimate places India's enrichment capacity at 20 kg/yr of HEU (93 per cent U-235).[20] Such a meagre yield, however, is deemed to be barely sufficient a fuel source for the nuclear reactor being developed for India's nuclear submarine, leaving no significant amounts for HEU-based weapons production. Some research activities in laser enrichment also are being undertaken, but they too are inconsequential by way of being a major source of fissile material.

The only other material that can be used for weapons by India

16. Commissioned in 1985, Dhruva became fully operational only in 1988.
17. It has been operating since 1985.
18. It became operational in 1990.
19. Gregory S. Jones, *From Testing to Deploying Nuclear Forces: The Hard Choices Facing India and Pakistan* (Santa Monica: RAND, 2000), p. 8.
20. Zia Mian, A.H. Nayyar, R. Rajaraman and M.V. Ramana, *Fissile Materials in South Asia and the Implications of the US-India Nuclear Deal,* Draft Report for the International Panel on Fissile Materials (IPFM), July 2006, p. 17. The same study on p.18 estimates that Pakistan has the capacity to enrich 100 kg/yr of uranium. It is already expected to have a stockpile of about 1,300 kg of 90 per cent enriched uranium, sufficient for 65 weapons at 20 kg HEU per weapon. Henceforth IPFM Study.

is its stockpile of reactor grade plutonium (RG Pu). It is estimated that at present, India could theoretically have about 11.5 tons of plutonium in its spent fuel obtained from unsafeguarded PHWRs.[21] These, however, are only indicative estimates based on certain assumptions about the efficiency and capacity factors of power plants. Naturally if the value of the assumption changes, so would the figure. Moreover, even if the calculated reactor grade stockpile is taken to be valid, how much of this has undergone plutonium separation is not known. Although India is believed to have used RG Pu in a low yield explosive device tested in May 1998, scientists the world over are divided over the use of this material for weapons. Some maintain that there are several disadvantages of crafting bombs from RG Pu, owing to its high content of Pu-240, that makes it spontaneously fissionable, and, hence, not only difficult to handle but also unreliable. On the other hand, there are studies to indicate that countries with advanced nuclear technology can use RG Pu effectively, and the UK is believed to have done so.

Nevertheless, the likelihood of India using its RG Pu stockpile for weapons is constrained by the fact that it needs these stocks (@ 3-4 tons per breeder) [22] to fuel the FBR that comprises the second stage of its 3-stage nuclear power programme.[23] The first of these (a total of five are planned as of now) is under construction at Kalpakkam, Tamil Nadu and it would have to be fuelled with indigenous sources of RG Pu. The fuel that is used in an FBR is a

---

21. Ibid., p.15. Spent fuel obtained from safeguarded reactors (TAPS 1 and 2) has not been reprocessed and is stored in spent fuel pools though India is under no legal obligation to not reprocess it.
22. R. Ramachandran, "Is Breeder Needed for Strategic Purposes?" *The Hindu*, February 22, 2006.
23. As envisioned by Homi Bhabha nearly half a century ago, India has opted to develop a three-stage fuel cycle to gradually reduce its reliance on uranium, of which it has only 0.8 per cent of world reserves, and move to the use thorium, of which India has 32 per cent of world reserves. For a detailed description of the three stages, see DAE website, http://www.dae.gov.in

mixture of depleted uranium (mostly non-fissile U-238) and RG Pu in the ratio of 7:3 or more. But after a doubling time that could range from 9-19 years[24], depending on whether the uranium-plutonium mixture is in metallic, nitride, carbide or oxide form, the FBR would produce weapons grade plutonium (WG Pu) and U-233.[25] Both these are fissile materials that could be used to produce weapons, though in the Indian case, they have a role in fuelling the third stage of its power programme wherein Advanced Heavy Water Reactors (AHWRs) will use U-233 along with indigenously available deposits of thorium to produce large amounts of energy for the country. Even if some proportion of the material produced in the FBRs is extracted for strategic purposes, the point is that this is likely to be available for such use only by 2020 or thereabouts, by which time the FMCT could be well in place making it impossible for India to use this material for weapons because it would have come under safeguards.

Therefore, for the moment, Pu-239, as obtained from CIRUS and Dhruva, is the primary fissile material for weapon production available to India. Varied estimates about this inventory have long been circulated in Western literature. Interestingly, nearly 14 years before India declared itself a state with nuclear weapons, Arnold Kramish, in 1984, had predicted that by the year 2000, India would possess a staggering 8,800 kg of Pu.[26] He based this estimate on the premise that India would use all its power reactors for weapons grade plutonium production. Also about the same time, a Congressional Research Service report estimate placed

24. Ramachandran, n. 22. However, according to Dr Kakodkar, Chairman DAE, the doubling time could be about 10-12 years.
25. Nuclear fuels in metallic form have relatively shorter doubling times. But the present PFBR will have oxide fuel.
26. Arnold Kramish, "The Bombs of Balnibarbi," in Rodney Jones, ed., *Small Nuclear Forces and US Security Policy* (Lexington, KY: Lexington Books, 1984), pp. 26-30.

India's Pu stockpile at a much lower 1,600 kg by 2000.[27] Soon after India's nuclear tests, David Albright of ISIS reported that India had about 370 kg of weapons grade plutonium, which according to him translated into 74 nuclear weapons.[28] About the same time, *Jane's Intelligence Review* estimated that India's nuclear stockpile contained between 20-60 warheads.[29] George Perkovich, in his book published in 1999, estimated India's Pu stockpile at about 250-300 kg in 1998 and found it sufficient for about 50 weapons.[30] Indian writers, too, have corroborated this conservative figure. For instance, R. Ramachandran calculated the WGPu stockpile to be not more than 280 kg in 1998 given the assessment of efficiency and capacity factors of CIRUS and Dhruva, as also of the reprocessing facilities.[31] According to a more recent report on this subject, it is estimated that India presently has about 500 kg of WG Pu that would give it about 100 weapons.[32]

This may or may not be true given that the "fissile material-to-number of weapons" equation is not a direct derivation since there are numerous other variables that come into play. Depending on the quality of its weapon design and the type of weapon to be produced (fission device or thermonuclear weapon), the number of nuclear warheads that any given stockpile could translate into would vary significantly. In fact, a

27. Study as cited in Bharat Karnad, "Going Thermonuclear: Why, With What Forces, At What Cost," *USI Journal*, July-September 1998, p. 315.
28. David Albright, *Fact Sheet: India and Pakistan – Current and Potential Nuclear Arsenals* (Washington DC: Institute for Science and International Security, May 1998).
29. W.P.S. Sidhu, "India Sees Safety in Nuclear Triad and Second Strike Potential," *Jane's Intelligence Review*, July 1998, pp 22-25.
30. George Perkovich, *India's Nuclear Bomb* (Berkeley, CA: University of California Press, 1999).
31. R. Ramachandran, "Pokhran II: The Scientific Dimensions", in Amitabh Mattoo, ed., *India's Nuclear Deterrent* (New Delhi: Har Anand Publications, 1999).
32. IPFM Study, n. 20.

given quantity of fissile material could be used to produce more or fewer number of weapons. For example, if India's present stockpile is taken to be 500 kg, then it could be used to fashion 166 weapons of 3 kg Pu, or 100 weapons of 5kg Pu each or 62 weapons of 8 kg Pu each. International Atomic Energy Agency (IAEA) regulations assume that 8 kg Pu would be required to produce a simple fission device yield of about 20 kt .[33] However, some scientists have argued that this is an unnecessarily large amount of Pu and that depending on the technical capability of countries, the same yield could be achieved with much less Pu. Countries with high technical capability could reach the desired result with only 3 kg. In the case of India, while some believe it to have the capability to make 20 kt weapons with about 5 kg weapons grade plutonium[34], others surmise the use of 6-8 kg WG Pu for a 12-15 kt yield.[35]

The other significant variable in this context is the distinction between fissile material production capacity and the actual quantities of fissile material that are available at a given time for weapons production. For instance, CIRUS and Dhruva might theoretically be capable of producing about 32 kg/yr of Pu[36], but the actual fissile material for weapons can only be obtained after reprocessing of this fuel at Trombay, where India's only reprocessing facility for WG Pu exists, and that would depend on the capacity factor and efficiency levels of the reprocessing plant. Also, besides the availability of Pu-239, weapons production would also need sufficient quantities of deuterium, tritium, lithium deuteride and other special

33. As quoted in Tellis, n. 14, p. 486.
34. Ibid. Also see IPFM Study, n.20.
35. Ramachandran, n.22.
36. It has been estimated that at full power and 70 per cent capacity factors, CIRUS could produce 9.2 kg/yr of WG Pu and Dhruva 23 kg/yr. IPFM Study, n. 20, p. 12.

materials such as polonium and beryllium. Therefore, it is actually a misrepresentation of facts to derive the number of nuclear weapons directly from the fissile material stockpile. This can at best provide only an indicative figure, which may or may not correspond to the real number of weapons available with the country at any given moment.

Conclusion

There is little doubt that the FMCT would put a stop to the future stockpiling capabilities of India. However, going by the estimates available on the likely stockpiles that India may already have, or will be able to accumulate before an FMCT comes into force, it may not be completely off the mark to believe that this could qualify for credible deterrence. While it may be true that the "minimum" needed for deterring Pakistan cannot be the same as that for China or another nuclear superpower, the point that merits constant reiteration is that *deterrence is not proportionately derived only from the fissile material or the number of nuclear weapons*. It has to be calculated on the basis of a more complex formula that takes many other factors into account. Fissile material is a critical component of deterrence. Yet, it is but one of many and cannot guarantee deterrence by itself.

In fact, the credibility of deterrence is determined more by the weapons that can *survive* a first hostile strike (given India's no first use posture) to render *assured retaliation* that will wreak *unacceptable damage* on the adversary. The key issues here are;

(a) the survivability of nuclear warheads, delivery systems, command and control and communication networks;

(b) the availability of reliable and accurate delivery mechanisms of the requisite range;

(c) the will to undertake assured retaliation;

(d) a pre-determination of what constitutes unacceptable damage to the adversary, so that it reconsiders, and aborts any decision to strike at India in the first place.

Seen thus, it appears unnecessary to argue for an open-ended build-up of fissile material stockpile to match or surpass that of the adversary. This would, in fact, only add to India's command and control and security costs and complexities. Instead, India would do better to concentrate on improving the range and precision of its delivery mechanisms and increasing the survivability quotient of its nuclear forces (which would include not just the warhead and its delivery system, but also the necessary C3I, targeting strategies and even its civil defence infrastructure) to further enhance deterrence.

9

Tale of the Twins
Overlapping Dimensions of India's Nuclear Energy and Strategic Programmes

Given the inherent duality in the nature of nuclear power – civilian exploitation of the atom for electricity generation and military use of fission for weapons – theoretically there could be overlapping dimensions of every nuclear programme. However, in most NNWS where nuclear activities are under full scope safeguards (FSS), there is some assurance about the peaceful nature of the nuclear programme. In the NWS, on the other hand, this duality matters little since they are anyway known to have military nuclear programmes. In the case of India, however, which is neither an NWS nor an NNWS (as defined by the NPT), the coexistence of the nuclear energy and strategic programme posed a challenge, especially when India sought to engage in international nuclear commerce that is predominantly governed by an elaborate set of guidelines drawn up by the Nuclear Suppliers Group (NSG), a club of nuclear supplier nations, which was born as a direct outcome of India's peaceful nuclear explosion (PNE) in 1974, and which has no provision for dealing with a non-NPT state.

From about 2003-04 onwards, India had begun to argue for its "exceptionalisation" from NSG guidelines in deference to its *unique needs* (large energy requirements that would best be met through a low carbon energy source) and *unique status* (impeccable non-proliferation record).[1] Given the inability of the NPT to accommodate India as a *de jure* NWS and the impossibility of denying India's *de facto* nuclear weapons capability, it was evident that a way had to be found to bring India, an advanced nuclear power, into the non-proliferation regime.

The US Administration under President Bush took up this challenge and the Indo-US Joint Statement of 2005 offered a way out. It envisaged cooperation with India's civilian nuclear energy programme provided it could be effectively fire-walled from its strategic dimension. Implicit in this offer was the recognition that India has, and would sustain, its nuclear weapons programme, only that the paths of the two programmes would not cross. Therefore, in an unprecedented gesture, civilian nuclear cooperation was offered without India having to accept full scope safeguards, a long held prerequisite for nuclear cooperation under US laws and NSG guidelines.

Given the uniqueness of India's accommodation (where it was being allowed to retain its weapons programme) and the far ranging procedural amendments necessary (such as waivers or legislations), it was hardly surprising that political leaders, academics, strategic analysts and even the media, not only in the USA and India, but also elsewhere across the globe, unspooled a range of concerns. Much has been written on these since July 2005 and it is not the intention here to document the debate that enveloped India's claim for access to international nuclear commerce. Instead, the purpose of this

1. For an articulation of this viewpoint, see Manpreet Sethi, "NSG Guidelines: Need for a Country-Specific Approach," in Jasjit Singh, ed., *Nuclear Power and Proliferation: Conflict or Convergence* (New Delhi: Knowledge World, 2004), pp.29-50.

exercise is two-fold: one, to understand the distinct inter-linkages between the Indian nuclear energy and strategic programme. This examination is critical because given the circumstances in which India's nuclear activities evolved, the civilian and military dimensions of its nuclear programme are more overlapping than in other nations. The bonds between the two are also a result of India's unique three-stage nuclear power strategy. The first stage involves the use of uranium as fuel in PHWRs and the reprocessing of spent fuel to extract Pu. In the second stage, Pu is used in the nuclear cores of FBRs, and surrounded by a "blanket" of either uranium (depleted) or thorium to produce more Pu or U-233. The third stage involves breeder reactors using U-233 in their cores and thorium in their blankets. This fuel cycle for generation of electricity, however, also produces fissile material (Pu) for India's nuclear arsenal. Hence, some of the processes are common and the same nuclear facilities, particularly the reprocessing plants, feed both the programmes. However, the co-habitation of military and civilian nuclear facilities makes it impossible to receive nuclear cooperation because the very possibility that assistance to the peaceful programme could indirectly end up aiding the weapons programme is anathema to the international non-proliferation community. Therefore, separation of the civilian and nuclear dimensions of India's nuclear programmes and acceptance of international safeguards on facilities declared civilian is essential for cooperation.

The second purpose of this chapter is to examine the implications of this cooperation, and any policy changes that it may require, on India's military programme. Though the determination of what would be open to safeguards and what would be retained in the military domain is an Indian decision, many have questioned the very need to create two separate nuclear programmes when the fuel cycles are so closely intertwined. Why duplicate facilities and bear additional costs? Why do we need international nuclear cooperation at all? And would this cooperation, with its concomitant compromises, not adversely impact the strategic programme? To this effect, the first section of this chapter explores the close inter-linkages between the energy and strategic dimensions of India's nuclear programme and how the fortunes of one have affected the other. It also explains why separation is necessary and assesses the problems that could arise in the process.

The second section evaluates the merits of rapid expansion of nuclear energy for India's energy security and, hence, underscores the benefit of cooperation. It also addresses the concern over fuel supply assurances because unless reliable, secure import of uranium is available to run the power plants, investment in a large number of them would not only be meaningless but also wasteful.

The issue of reliability of uranium supplies then leads to the third section of the chapter that examines two specific implications of this cooperation on the country's strategic programme: the ability to conduct nuclear tests; and to accumulate requisite fissile material for credible minimum deterrence. The chapter concludes by suggesting that ensuring energy security in a mature and pragmatic fashion is as much a strategic necessity for the nation as its nuclear weapons programme. It would be foolish to ignore an opportunity that holds the promise of bridging India's electricity deficit in an environmentally sustainable manner for the sake of keeping open some notional options of nuclear testing and fissile material accumulation, particularly since the nature of India's nuclear deterrence does not really deem it necessary.

The Linkages

Every one of the five NWS pursued a dedicated nuclear weapons programme before they turned their attention to exploiting the peaceful aspects of nuclear energy. However, unlike any of them, India's nuclear weapons programme grew out of its civilian nuclear energy efforts. In fact, it was on March 12, 1944, even before the use of the atomic bombs that Homi J. Bhabha had written to the trustees of Sir Dorabjee Tata Trust proposing the establishment of an institute to train nuclear scientists. He wrote, "When nuclear energy has been successfully applied for power production, in say a couple of decades from now, India will not have to look abroad for its experts, but will find them ready at hand."[2] Of course, this does not suggest that the two early protagonists of the Indian nuclear programme, Bhabha and Jawaharlal Nehru, were unaware of the utility of nuclear weapons as a new currency of national power. In fact, in his presidential address to the Indian Science Congress in 1947, Nehru acknowledged that atomic energy "may be used for war or may be used for peace. We cannot neglect it because it may be used for war... we shall develop it, I hope, in cooperation with the rest of the world and for peaceful purposes."[3] Therefore, the focus was largely to tap the civilian potential of the atom, though Bhabha recognised that the indigenous pool of nuclear scientists and technologies developed over time could equally be used for making the weapon, if required. Accordingly, India legislated the Atomic Energy Act on April 15, 1948, leading to the creation of the Atomic Energy Commission on August 10 the same year.

It may be recalled that internationally too this was the period of nuclear euphoria[4] when people believed that nuclear electricity would be so cheaply produced that it would not require to be

2. H.N. Sethna, *Atomic Energy* (New Delhi: Publications Division, 1972), p.1.
3. As quoted by Itty Abraham, *The Making of the Indian Atomic Bomb: Science, Secrecy and the Postcolonial State* (New Delhi: Orient Longman, 1999), p. 47.
4. For instance, David Dietz, an American journalist and Pulitzer prize winner wrote, "With energy as abundant as air we breathe, there will be no longer any reason to fight for oil or coal..." in his *Atomic Energy in the Coming Era* (Dodd Mead: 1945), pp. 12-23. Glenn Seaborg , adviser to the US Atomic Energy Commission in the 1950s too described it as a "magician's potion that could free industrial society permanently from all practical bounds," in Seaborg and William Corliss, *Man and Atom: Building a New World through Nuclear Technology* (Dutton, 1971).

metered. This phase coincided with India's own desire soon after independence to embark upon a pioneering scientific field in order to effectively mark its arrival on the international scene after 200 years of colonial rule. President Eisenhower's Atoms for Peace programme in 1953, therefore, proved to be timely, as was Bhabha's chairmanship of the International Conference on Peaceful Uses of Nuclear Energy in 1955. In his opening address, he highlighted the importance of this energy for developing nations: "For the full industrialisation of the underdeveloped countries and for the continuance of our civilisation and its further development, *atomic energy is not merely an aid, it is an absolute necessity*."[5]

Making use of his contacts abroad, Bhabha secured nuclear cooperation for India from a number of sources. In June 1954, he requested Sir John Cockroft, his colleague from Cambridge and an important figure in the British atomic programme, to help India build a low power research reactor. With the help of detailed engineering drawings, technical data, and enriched uranium fuel rods for a "swimming pool reactor," Apsara, became critical in August 1956. The second research reactor to attain criticality, in 1960, was CIRUS – a 40 MW heavy water moderated (which was supplied by the USA), light water cooled, natural uranium fuelled reactor supplied by Canada, the country that also helped India set up its first power reactor, a PHWR at Rawat Bhatta in Rajasthan. Meanwhile, the US helped India construct two 200 MWe (later 160 MWe) Boiling Water Reactors (BWR) at Tarapur. Built through a turnkey project, Tarapur Atomic Power Station (TAPS) went critical in 1969 and provided India with valuable reactor construction and operating expertise, besides electricity to the grid.

The bilateral nuclear relationship between India and the

5. United Nations, *First International Conference on the Peaceful Uses of Atomic Energy* (New York, 1955), vol. 16, p. 33. Emphasis added.

USA, however, soured once India conducted a PNE in 1974 and the period of nuclear estrangement lasted until the July 2005 Joint Statement heralded the possibility of a new engagement. Meanwhile, Tarapur resonates for different reasons in the nuclear memories of both nations. For the US, it signifies India's betrayal of trust since Washington believes that India had used the heavy water supplied for CIRUS and the plutonium produced therefrom in its nuclear explosive device. Hence, under US laws, it ceased all cooperation with India and also reneged on its contractual obligations to supply enriched uranium to fuel the two power plants at Tarapur. This experience weighs heavily on the Indian mind as it seeks to reengage with the US on nuclear commerce. Especially so, since India maintains that it violated no contractual or other commitments by conducting the peaceful nuclear explosion.[6]

Coming under sanctions and denied access to all kinds of dual use technology, the list for which went on expanding through the 1980s and 1990s, India's nuclear power programme was forced after 1974 to primarily bank on indigenous R&D and domestic industrial efforts. This, of course, resulted in time delays and cost overruns for existing projects. Installed capacity in 1979-80 was about 600 MWe, and it could climb to no more than 950 MWe by 1987. In fact, after 1973 when Rajasthan Atomic Power Station (RAPS) 1 went on line, until 1981 when RAPS 2 started commercial power production, there was a long gap. Only two other power plants, Madras Atomic Power Station (MAPS) 1 and 2, became

6. The conduct of PNEs, during the 1960s and 1970s, was considered a legitimate economic activity for civil engineering purposes. The US and USSR, of course, conducted several PNEs. Germany too, in the early 1970s conducted a feasibility study for a project to build a canal from the Mediterranean Sea to the Western Desert of Egypt, using nuclear demolition. This project proposed to use 213 devices, with yields of 1 to 1.5 megatons detonated at depths of 100 to 500 m, to build the canal for the purpose of producing hydroelectric power.

critical in the 1980s. Four more – Narora Atomic Power Station (NAPS) 1 and 2 and Kakrapar Atomic Power Station (KAPS) 1 and 2 came on line in the 1990s so that by 2000, the total nuclear energy generation stood at 2,720 MWe.

Therefore, the 1974 PNE had a profound impact on the nuclear power programme. On the one hand, it compelled India to rely on its own scientific and technological resources. On the other, the halt to nuclear cooperation so embittered the Indian nuclear and strategic community that even 35 years later, the 'Tarapur experience' haunts any initiative to embark on a new nuclear cooperation arrangement with the USA, lest Washington unilaterally snap ties any time in the future in response to an Indian action that it perceives in violation of its domestic legislation and forces others to follow suit. It is for this reason that concern over securing "assured fuel supplies" for lifetime operation of imported reactors emerged as a major issue in the negotiations between New Delhi and Washington and is dealt with in greater detail subsequently in this chapter.

Meanwhile, staying with the 1974 incident, while the PNE disturbed the pace of the power programme, it was not particularly effectively encashed for any strategic gains. It certainly brought India on the nuclear proliferation radar and made it a victim of technology denial regimes, many of which were created as a consequence of the Indian action. Even as the power programme struggled, R&D for exploiting the military potential of the atom simultaneously, though secretly, continued. It grew in spurts, the momentum driven by intelligence reports received about nuclear proliferation in the neighbourhood. Finally, given the developing constricting nuclear environment around the mid-1990s, once the NPT was extended indefinitely in 1995 and the CTBT was opened for signature in 1996, India found it necessary to overtly demonstrate its weapons capability to establish nuclear deterrence.[7]

Not surprisingly, the country once again invited wide-ranging international sanctions. This time, however, owing to the largely homegrown nature of the civilian nuclear programme, it did not suffer as much. In fact, it may be recalled that by the second half of the 1990s, the indigenous nuclear power programme had fairly established itself and was generating 2,720 MWe of electricity. Seven more plants were under construction and the pace of work on power reactors continued largely unaffected. In fact, the constraint that began to be felt in the early years of the new millennium was not in nuclear technology, expertise or financing, but in availability of uranium to fuel the fast expanding programme of power reactors. This challenge, and the desire of the Department of Atomic Energy (DAE) to rapidly enhance nuclear power production through induction of additional imported, larger capacity power reactors, persuaded the government of the day to explore options for international civilian nuclear cooperation.

A window of opportunity presented itself when President Bush offered the promise of a constructive nuclear engagement with India. In a substantial *volte-face* compared to the policies of his predecessor, the new vision of President Bush was encapsulated in the joint Indo-US statement of July 18, 2005, signed when Prime Minister Manmohan Singh visited Washington. In this was implicit the recognition of India as a rising economic power with substantial energy requirements, and as a "responsible state with advanced nuclear technology." The other significant, though tacit, acknowledgment was of India as a nuclear weapons capable state that would not, and could not, be disarmed unilaterally. Therefore, from being viewed as an outcast to be chastised for "illegal"

7. For the range of factors that prompted India to move towards the 1998 nuclear tests, see Jasjit Singh, *Nuclear India* (New Delhi: Knowledge World, 1998).

possession of this WMD and to be coerced to give up its weapons, India's strategic programme was being recognised and accepted. The Indian Prime Minister confirmed this in his statement before the Parliament: "The existence of our strategic programme is being acknowledged even while we are being invited to become a full partner in international civil nuclear energy cooperation."[8]

In order to make use of this opportunity, India had to create a clear division between the peaceful and military dimensions of its nuclear programme. A Separation Plan was drawn up by the government in collaboration with the DAE to undertake this exercise in a phased manner and presented to the Parliament in March 2006. While an exact quantification of the cost of such an exercise has not been officially revealed, that there will be some extra burden is to be expected since some additional facilities, either in the civilian or military sector will need to be built. However, since the Indian nuclear programme is anyway in a phase of expansion, and the separation would be implemented over a period of time up to 2014, the cost will be spread over some years. Separation will turn out to be a useful activity in the long run because as and when the FMCT comes into place, this separation would be useful in the implementation of safeguards and international verification. Also, it would help bring greater transparency and accountability into the nuclear programme, and, hence, enhance its public acceptability. So, if nuclear energy is to be an important constituent of India's energy mix, and the following section explains why it should be, then civilian nuclear cooperation enabled by a separation plan must be undertaken.

India's Need for Nuclear Energy

Ever since the Industrial Revolution, energy, especially electricity,

8 Prime Minister's statement in Parliament on February 27, 2006. Full text available in *The Hindu*, February 28, 2006.

has been considered a fundamental factor in economic development. P.M.S. Blackett, a prominent participant in the British atomic energy programme had extrapolated in 1948, "It is not always recognized how very intimately a modern nation's industrial strength, and so both its standard of life and its military potential depend on the availability of power."[9] Since then, per capita Gross National Product (GNP) is taken to be roughly proportional to per capita energy consumption.

Economic growth and human development are the twin aspirations of an emergent India. But, among other factors, future economic growth will be contingent upon the long-term availability of energy, particularly in the form of electricity. For a country of India's size, population and economy, targeting ambitious growth rates over the next two decades, safeguarding energy is a strategic priority. This will have to be available in increasing quantities and from sources that are stable, dependable, low-cost and environmentally friendly. But, presently India imports the bulk of its energy requirements of coal, oil and gas, and, hence, the urgency to secure sources of supply. In an attempt at diversifying its energy mix, India has developed renewable and other alternative sources of energy, including nuclear. It is the latter which is the focus of worldwide attention today. A global nuclear renaissance is well timed with India's desire for nuclear expansion and this source has tremendous potential for a power hungry India.

Growing Energy Demand

World Bank estimates, broadly corroborated by a draft report prepared by the Expert Committee of the Planning Commission on Integrated Energy Policy, place growth rate of energy

---

9 P.M.S. Blackett, *Military and Political Consequences of Atomic Energy* (London: Jurnstile Press, 1948), p. 90.

demand in India at 5.3 per cent per annum in the first decade of the 21st century and by 2020, it is expected to grow to 10 per cent per annum.[10] Underpinning this trend will be the ongoing growth in population, urbanisation, income levels, industrial production and transportation demand. To meet such targets, it has been estimated that by 2031-32, electricity generation must grow from the present 633 billion kilowatt hours (bkWh) to 3,880 bkWh. The present total power generation of about 150 gigawatt (GW) is woefully short of current demand. Per capita energy consumption in India at present is at 600 kilowatt hours (kWh) as compared to 1,300 kWh in China and 15,000 kWh in developed economies. According to Dr Kakodkar, Chairman, Atomic Energy Commission, even if India's per capita energy consumption was to increase to 5,000 kWh (which would still be three times less than the current US consumption), the country would suffer an energy deficit of 412 GW by 2050. The deficit itself would be three times the current total production. As the economy becomes even more technology dependent in the coming years, electricity would play an even greater role as an economic multiplier. In case this demand is not met, the loss to India's economy, though difficult to quantify in figures, would be huge. Some contend that if the energy sector in India could reach international levels of performance, the country's Gross Domestic Product (GDP) would go up by an additional 2 per cent per annum.[11] Therefore, increasing energy generation to meet growing demand is critical for continued economic growth and social development.

---

10. Draft Report of the Expert Committee on Integrated Energy Policy prepared under the chairmanship of Dr Kirit Parikh, member (Energy) Planning Commission, Government of India. Full report available at website of Observer Research Foundation.
11. R.K. Pachauri, "Defining an Integrated Energy Strategy for India," TERI website at http://www.teriin.org/energy/

Limited Domestic Fuel Sources

The crunch, however, lies in the availability of energy sources. In the present Indian energy mix, coal is the dominant fuel meeting 55 per cent of primary energy consumption, with the share of oil at 36 per cent of total energy consumption. Although India's coal reserves comprise 8 per cent of the world total and the country is its fourth largest producer, only 20 per cent of the total coal transported to power plants is of superior grade, with ash content of 24 per cent or less. The remaining about 80 per cent is of inferior grade, which when used in thermal plants, not only poses environmental problems, but also raises the cost of operation and maintenance, and adversely affects plant performance.[12] Over the years, therefore, imports of high quality coal have risen from 6 MT in 1991/92 to 20 MT in 1999/2000, a rate of growth of 16 per cent a year.[13]

The country's hydrocarbon reserve is also small at 0.4 per cent of world reserve. In fact, current reserve-to-production ratios for coal, oil and natural gas available in India are 235, 23 and 35 years respectively.[14] By 2030, India's demand for oil will require imports of almost five million barrels a day, unless there is a substantial increase in domestic petroleum finds and exploitation. This will increase India's import dependence for crude oil from 70 per cent in 1999-2000 to 90 per cent in the coming decades.[15] Natural gas is emerging as a replacement for oil, but this too will have to be imported. Indian demand for natural gas is expected to grow by over 5 per cent per annum so as to meet 13 per cent of primary

12. Sajal Ghosh, "Sustainable Energy Policies for Clean Air In India," *Background Paper* prepared for The Atlantic Council, USA, p. 9.
13. Ministry of Commerce, *Annual Report* (New Delhi: Ministry of Commerce, 2001).
14. Ghosh, n. 12, p.9.
15. Jasjit Singh, ed., *Oil and Gas in India's Security* (New Delhi: Knowledge World, July 2000), p.ix.

demand, but this too will be met by 70 per cent imports.[16] This scenario is alarming not only because of the pressure it places on scarce foreign exchange resources, but also because it raises energy vulnerabilities, a fact that for a large country like India is neither affordable, nor strategically prudent.

Growing Global Environmental Concerns

When the bulk of the country's energy is sourced through fossil fuels that disperse CO_2 directly into the air, environmental consequences like greenhouse effect and global warming become a serious cause for concern. Thermal power plants pose the problem of greenhouse gas (GHG) emissions that cannot be wished away despite technology improvements and application of stringent environmental measures. Rather, the pollution is bound to increase with the upsurge in energy production from thermal plants and would have climatic implications. In order to arrest these disturbing trends, environmental experts warn that the overall global CO_2 emissions should be cut from 25 billion tons annually to 10 billion tons, even as overall energy production increases – not a small challenge by any standards.

At present, India's per capita energy consumption and, hence, the resultant carbon emissions are relatively low. In 1997-98, the average per capita energy consumption of India was less than 0.5 tons of oil equivalent (TOE) compared to the world average of 1.7 TOE.[17] Consequently, per capita carbon emissions of India in 1997 were only 250 kg of carbon per person, which was approximately one-quarter of the world average and less than 0.05 per cent

16. Jasjit Singh, "Growing South Asian Interests in the Persian Gulf Region: Problems and Opportunities," *Strategic Analysis*, vol.23, No.9, December 1999.
17. Amit Kumar,"Energy Scenario in South Asia," *Energy Technology News* (TERI), Issue 1, October 2000.

of that in the US[18]. Despite these low figures, India is still the world's sixth largest carbon emitter. This ranking would certainly change in the coming years as India emerges as the world's most populated and urbanised country by 2025, and even more so if it continues to retain coal as the predominant fuel in its energy mix. It is estimated that India's total carbon dioxide emissions will rise from the present 0.9 billion tons to 2.3 billion tons by 2030, making it one of the largest carbon emitters.[19]

Given these estimates, the strategies and technologies India adopts to meet its energy demand will have critical implications for its own, and the global environment. Nuclear power, in this scenario, could prove helpful since it has been widely accepted as a zero carbon source. In fact, this attraction of nuclear power would further be enhanced once the system of carbon emissions trading is perfected because a price on carbon emissions would provide a financial incentive to reduce the carbon footprint. The European Union has implemented Phase I of the emissions trading scheme and the next phase is to be enforced from 2008 onwards. While the system is not functioning perfectly yet, attempts are being made to fine-tune it. The bottom-line, however, remains that once governments set a high price on carbon polluting electricity, it will make investment in carbon free electricity technologies, like nuclear power, more attractive.

India's Unique Benefits from Nuclear Power

With experience of half a century in the nuclear field, India is today capable of designing, building, operating and maintaining nuclear power plants of varied types and capacities[20], besides being able

18. Ibid.
19. World Nuclear Association, http://www.world-nuclear.org
20. India operates nuclear plants of types as different as Boiling Water Reactors of 160 MWe, Pressurised Heavy Water Reactors of 220 and 550 MWe, and is constructing the Light Water Reactors of 1,000 MWe with Russian help. It

to manufacture all associated equipment and components, and produce the required nuclear fuel and special materials. Having amassed substantive reactor experience in PHWR operations, India is moving towards the commercial demonstration of the fast breeder reactor programme. The Prototype Fast Breeder Reactor (PFBR) is being constructed at Kalpakkam and is likely to come on line by 2010. At the same time, the country is developing the reactor and associated fuel cycle technologies for thorium utilisation, with a one of its kind 30 KW(Th) research reactor KAMINI already operational. Of course, the thorium cycle faces problems of high cost and technical complications in fuel fabrication and reprocessing because of the high radioactivity of the materials involved. But then, India is the only country pursuing this technology. Even the World Nuclear Association, which is dedicated to the promotion of nuclear technology, sees little future in it while abundant uranium is available. But uranium availability is a major constraint for India.

The Indian uranium reserves, at about 0.8 per cent of the world's are considered to be insufficient for a programme beyond 10,000 MWe if uranium is used on a once-through basis. India has, therefore, undertaken spent fuel reprocessing to supplement its nuclear fuel resources with recovered Pu. In a three-stage programme, the available uranium will eventually be used to harness the energy contained in non-fissile thorium, of which India possesses about 32 per cent of the world's reserves or 360,000 tons of high quality thorium. But even best estimates suggest that the promise of thorium can be realised only by 2050. Until then, if India has to make optimum use of its nuclear experience and expertise and to accelerate its nuclear generating capacity, it must access imported nuclear fuel. India's proven domestic uranium reserves

---

is also constructing the 500 MWe Prototype Fast Breeder Reactor as well as developing the Advanced Heavy Water Reactor for thorium use.

are mostly of low-grade quality and entail a high cost of extraction. Moreover, due to the lack of focus and/or public opposition over the last several years to uranium prospecting, mining and addition of milling capacities, the operation of the existing nuclear reactors is at present severely handicapped due to scarcity of fuel. Therefore, even as plant construction is proceeding at a good pace, there is not enough nuclear fuel available to operate the reactors at optimal capacity levels. In fact, it is reported by the NPCIL itself that it has had to downgrade the plant load factors (PLF) in several reactors to keep them running. Consequently, there has been a rapid drop in capacity utilisation factors from 85-90 per cent only four years ago to 60% by August 2007[21]. By February 2008, the PLF of MAPS 1 had further slid to 36 per cent from 72 per cent and that of K APS I to 46 per cent from 67 per cent in 2007.[22]

Of course, this slide, would be arrested once new mines and milling plants are opened or as nuclear fuel is imported from 2009 onwards with the lifting of the constraints that had denied India's participation in international nuclear commerce, and the conclusion of bilateral nuclear cooperation agreements with countries, India is looking to import nuclear fuel for its indigenous nuclear power plants. Given that nuclear fuel and its continued and reliable availability will be at the heart of the expansion plan to ensure the smooth, uninterrupted operation of domestic and imported nuclear reactors, this issue needs to be dealt with in some detail.

Nuclear Fuel Supply Assurances

According to the Separation Plan drawn up by India in March 2006, it was clearly spelt out that in exchange for India's acceptance of

21. "Pressing Pause on Deal will Deepen Crisis at N-Power Plants," *Indian Express*, August 20, 2007.
22. Amitabh Sinha, "Low on Fuel and no Sign of Deal, Nuclear Power Plants Take a Hit," *Indian Express*, March 27, 2008.

safeguards in perpetuity on its declared civilian nuclear facilities, it would be assured lifetime supply of fuel for the safeguarded reactors. However, aspersions were cast on US intentions and sincerity in this regard when the Hyde Act passed by the US Congress in December 2006 replaced the promise of "lifetime supplies" with "reasonable reactor operating requirements."[23] This was largely pushed by the American non-proliferation lobby to reflect its own concerns. They were apprehensive that the easy and assured availability of lifetime supplies of imported fuel would tempt India to unhesitatingly use its domestic uranium for production of nuclear weapons. By granting India the assurance of lifetime supplies, it was feared that the US would lose all leverage in case India decided to conduct a nuclear test in the future.

The Hyde Act naturally invited a lot of criticism in India. However, the bilateral nuclear cooperation agreement or the "123 Agreement" (so-called for the Section 123 of the US Atomic Energy Act that pertains to bilateral civilian nuclear cooperation) concluded in July 2007, restored the fuel supply assurances that had been first made to India.[24] Article 5.6 of the agreement specifically mentioned that the US would help to create the necessary conditions for assured and full access to fuel through the amendment of its domestic laws, working with "friends" to adjust NSG practices, and with the IAEA for an India-specific fuel supply arrangement in order to help India build a strategic fuel reserve for lifetime of safeguarded reactors. These multi-layered fuel supply assurances were unique in international nuclear negotiations and included India's right to take "corrective measures" in case of disruptions. Significantly also, Article 14.8 insisted that the right of return, if invoked by the US, would not derogate from India's right for assured fuel

23. See text of Henry J. Hyde Act, Section 103(b)(10).
24. See Appendix 5 for text of 123 Agreement.

supplies from the international community even after the US terminates cooperation.

Despite the explicit inclusion of these clauses in the 123 Agreement, several concerns on fuel supply assurances have been voiced in India. Four of these merit a closer examination. The first stems from the apparent contradiction between the stances of the Hyde Act and the 123 Agreement. While the former promises no more than "reasonable reactor operating requirements," the latter provides assured and full access. How will the US be able to resolve this contradiction in case of a crisis? In this context, two aspects – one technical and the other political – need to be brought out. One, it must be understood that the Hyde Act is an enabling domestic legislation that has made the negotiation of the bilateral 123 Agreement possible since otherwise the US Atomic Energy Act of 1954 does not allow civilian nuclear cooperation with a country that has conducted a nuclear test, that does not accept full scope safeguards, and that is engaged in nuclear weapons related activities. Second, the issue of fuel supplies in the Hyde Act is mentioned in its section of "Statement of Policy" which is a non-binding and advisory segment of the document. In comparison, the relevant Article in the 123 Agreement forms part of operational law, hence, its sanctity should not be underestimated. Once the 123 Agreement has been approved by the US Congress and signed by the two governments (which was done in October 2008), it would govern the bilateral nuclear cooperation. Third, past experience indicates that enforcement of US laws is largely determined by the political relationship between nations. If the relations are healthy, laws have been ignored, twisted, played down or even amended to accommodate necessary actions. The Pressler Amendment and its selective enforcement vis-a-vis Pakistan is a clear example of this. When it suited American strategic interests to maintain robust relations with Pakistan, even the knowledge of its clandestine nuclear weapons programme and, hence, the enforcement of the Pressler Amendment that would have necessitated a complete cessation of military and economic assistance to the country was completely ignored.[25]

A second concern stems from doubts about US sincerity in helping India build a multilateral strategic fuel reserve. Given that there is a strong non-proliferation lobby in the US, with supporters across the globe, that is loath to surrendering any leverages over India, would a reserve that is sufficiently large be allowed to be built? In fact, President Bush's statements in September 2008 contradicted his earlier assurances in this regard. It must be admitted that India will not find it very easy to immediately build a large reserve. However, its ability to do so over time will depend more on its being able to negotiate bilateral contracts for the purchase of uranium or uranium prospecting and mining from individual supplier nations. A sustained effort at reserving resources, buying uranium mines or exploratory rights in other nations as well as accelerating uranium prospecting and mining within the country will be necessary. It must be remembered that with the NSG waiver, India's options of fuel imports (and other nuclear imports) significantly broaden. Simply put, it would enable India to have a wider choice of partners in nuclear commerce and technology trade. Meanwhile, as the American stake in the Indian economy increases and strategic cooperation strengthens, the US may not be averse to helping India build a strategic multilateral fuel reserve, or at least not obstruct it from doing so.

The third concern that has been voiced is whether with the import of nuclear fuel, India would not end up replacing one fuel

25. It is significant that the US ignored its own non-proliferation policies and laws to make Pakistan its frontline state to fight Communist Soviet presence in Afghanistan in the early 1980s because this was a precondition made by Pakistan to play this role. See Gen K.M. Arif, *Working with Zia: Pakistan's Power Politics 1977-88* (Karachi: Oxford University Press,1995),p.431.

dependency for another. While it is now dependent on coal, oil and gas for its energy needs, it would then become dependent on import of nuclear fuel. This apprehension, however, tends to overlook the fact that dependence on imported uranium would, at most, be transitional as India graduates into the second and third stages of its nuclear power programme that envisage the large scale use of the country's vast thorium reserves. India would use nuclear fuel imports to generate additional capacities from FBRs; and AHWRs would eventually make it self-sufficient in nuclear fuel. Therefore, unlike the dependence on import of fossil fuels that can only increase vulnerabilities exponentially as resources dwindle in the future and prices rise, import of uranium offers the hope of moving into energy self-sufficiency some time in the future.

The last concern arises from the possibility of supply disruptions. These may occur for a variety of reasons, but two of these are particularly important to consider. The first cause of disruption could be market disturbances as a result of the nuclear industry stepping into an overdrive, especially in Asia. While uranium demand is increasing, supplies might not be able to keep pace given that the slump in uranium production over the past few decades had led to low interest and investment in uranium mining. Prices of uranium had doubled in 2008, though by the end of the year that they were substantially lower.[26] And, it is for this reason that India, akin to what China is doing, would do well to build a strategic fuel reserve. It would be prudent for New Delhi to negotiate relevant imports from willing major uranium supplier countries such as Canada, Kazakhstan, Brazil, South Africa, etc. Possibilities of getting uranium from even non-NSG countries such as Niger where a private Indian entrepreneur has acquired

26. For more on this, see Sean Brodrick, "Seven Reasons the Uranium Price will Hit $100 this Year," http://www.moneyweek.com, August 21, 2007.

uranium-mining rights must also be explored and aggressively pursued.[27]

The second possible disruption of nuclear fuel could arise if India was to conduct a nuclear test. In such case, American laws mandate the cessation and termination of cooperation and the revocation of the "right of return" by the US President. Would the Indian nuclear programme then become hostage to US laws? The following section examines these implications in the context of India's nuclear strategy.

Nuclear Cooperation and India's Nuclear Strategy

Conduct of Nuclear Tests in the Future

The Hyde Act does make it clear that any future Indian test of a nuclear explosive device would invoke adverse legal and policy consequences. Under American law, the US President would be compelled to terminate all export and reexport of US-origin nuclear materials, nuclear equipment, and sensitive nuclear technology to India, besides demanding the return of nuclear-related items, materials, and sensitive nuclear technology that may have been exported or reexported to India. The termination of cooperation would also include the suspension and revocation of any current or pending export or reexport licenses.

The 123 Agreement, however, without specifically making any reference to "test and fuel supply disruptions" makes some allowances. Before any termination and cessation of cooperation, it allows both sides to "consider carefully the circumstances" in order to assess whether these "resulted from a Party's serious concern about a changed security environment or as a response to similar actions by other States which could impact national security."

27. "Indian Firm Acquires Uranium Mining Rights in Niger," *The Hindu*, August 19, 2007. Niger is the fifth largest supplier of uranium in the world.

The Agreement, therefore, does not provide for an automatic termination of cooperation, but a studied consideration of the situation that necessitated such a step. This is a huge concession since the US laws do not permit nuclear cooperation with a country that has tested a nuclear device but is not a member of the NPT.

Further, it is necessary to envision some of the conditions that could prompt India to conduct a nuclear test. One must not forget that when India accepted the voluntary moratorium upon itself after conducting five tests in May 1998, it was in the understanding that *further unprovoked nuclear testing* would not be required. If at all India may feel the need to test again, it would be in response to a perceptible change in its threat perceptions as a result of the conduct of similar activity by others. One such situation may arise in case India's nuclear deterrence with China is perceived to have been undermined after China's conduct of tests to buttress its nuclear capabilities. China's action would most likely be in response to a US move towards a concerted modernisation of its own nuclear weapons. Therefore, the international security environment would anyway have been vitiated enough for other countries to be conducting further nuclear tests. In such a scenario, if India felt compelled to carry out its own tests, would it really matter to the US? Would the Indian action then be seen as destabilising or stabilising? Of course, the law would necessitate a termination of cooperation, but given the circumstances, would the US President not be able to find a way around it? As mentioned earlier, politics plays a far greater role than legal technicalities in relations of such nature.

In fact, seen from another perspective, the clause in the 123 Agreement provides for an examination of Indian reasons for renewed testing and this amounts to placing China and Pakistan on notice that in case of an action by them that provokes India to test, the US could accept India's reaction as justifiable. Now, suppose India had tested in the absence of such an agreement, would it not have triggered off automatic American sanctions? With the agreement, New Delhi at least has the opportunity to explain and justify the reasons for its testing. Without it, no such understanding atmosphere would even be available. If anything, therefore, the text of the 123 Agreement accepts that India's national security might demand testing, a fact that had been unacceptable earlier. The creation of this sympathetic environment itself is a huge change from the fact that Washington, until not long ago, did not even acknowledge that India had legitimate security concerns that required the acquisition of nuclear weapons capability. In the ultimate analysis, of course, India's political relationship at that time with the US would have a strong bearing on its reaction to any Indian test. As mentioned earlier, if the relationship is strong, the US President could find ways around the legalese to continue cooperation.

Secondly, the 123 Agreement also stipulates the payment of compensation at market rates in the case of the US demanding a return of its nuclear imports. Besides the huge logistic and legal complexities involved in this exercise, surely it will not be seen as a welcome proposition by the American business lobby, whose stakes in the Indian economy would only grow over time. If anything, while drafting its nuclear contracts with foreign investors, India must seek to tie the nuclear companies into these terms and, thus, get them to act as a powerful lever in favour of continued cooperation.

The other essential point to consider is the very nature of India's nuclear deterrence as outlined in its nuclear doctrine, drafted well before such an agreement was anywhere on the horizon. Does the nature of India's nuclear deterrence require unprovoked qualitative improvements in order to be credible? Committing India to the political, deterrent use of the nuclear weapon, it presupposes the construction of a robust counter-strike capability that is survivable. It does not require India to indulge in a foolhardy exercise of bigger and more modern warhead construction through more and more testing. What India needs for credible deterrence is a proven design of a certain yield that can be reliably and accurately delivered at a pre-designated target to cause unacceptable damage. The credibility of India's nuclear deterrence, therefore, must rest on the reliability, accuracy, penetrability and survivability of its nuclear arsenal, primarily of its delivery capability. As long as a few missiles of the requisite range can reliably deliver nuclear warheads of even 15-20 kt yield on a modern urban conglomerate, it would cause damage that any rational state cannot but find unacceptable and, hence, be deterred.

Used in the modern mega cities, the scope for death and destruction of even the small weapons cannot be underestimated. In fact, scientific studies prove that if targeted at mega city centres, low yield weapons can produce 100 times as many fatalities and 100 times as much smoke from fires per kiloton yield as previously estimated in analyses of full scale nuclear wars using high yield weapons.[28] Clever targeting can further enhance the destructive potential of the same yield. Modern buildings with glass facades provide scope for more damage than traditional concrete structures. Similarly, the time of the day of the attack (on which the wind factor and human congestion of an area may depend), as well as the general lay of the land, influence the extent of destruction. In the case of Nagasaki, for instance, though the yield of Fat Man was higher than that of Little Boy, the number of dead was smaller because the hills surrounding the city shadowed areas

28. For more scientific and empirical details on this, see O.B. Toon et al, "Atmospheric Effects and Societal Consequences of Regional Scale Nuclear Conflicts and Acts of Individual Nuclear Terrorism," *Atmospheric Chemistry and Physics,* April 19, 2007, pp. 1973-2002.

from the blast and thermal radiation. Meanwhile, the fireball in the case of Hiroshima was more violent due to the dry conditions and flat terrain.

This reality must be kept in mind while assessing the impact of the Indo-US agreement on India's defence, rather than simply fearing the trading away of the right to test, which is not true either. India retains the sovereign option of testing, but at any time it takes such a step, it will have to be prepared for the consequences. However, it would be natural to assume that an economically stronger country, with a large American stake in the Indian economy would be better able to handle the consequences. Therefore, there was no reason for India to deny itself the opportunity to access international nuclear and other high technology commerce for its energy security and economic growth, only to keep the option of testing open. By doing so, India would have ended up undermining its economic strength, which too is a deterrent of another kind, while also losing the chance to reposition itself in the new world order.

Implications for Accumulation of Requisite Fissile Material

Another major fear in certain sections of the Indian strategic community arises from the acceptance of international safeguards over some of its nuclear facilities, thereby binding the country's hands on the number of nuclear weapons it can build. A discussion of the relationship between fissile material and India's definition of credible deterrence has already been undertaken in the last chapter. In this section, let me highlight two specific apprehensions related to India's acceptance of safeguards on some part of its nuclear programme.

One assumption that haunts Indian strategic thinkers is that the Indian nuclear doctrine of "no first use" demands a large number of nuclear weapons to deter the adversary. Since the country is not going to use this WMD first, but would have to absorb the first nuclear use, it is expected to need a large arsenal to assure some survival for retaliation. However, this is an erroneous assumption. It is not the size of the arsenal that assures survival, but other factors, as has been brought out in the chapter on survivability challenges in this book. There are several measures to enhance the survival of the arsenal that India must seek to buttress instead of solely focussing on fissile material accumulation for an expanded size of the arsenal.

The second fear stems from the argument of those who believe that for deterrence to be effective, fissile material inventories and weapon types of adversaries must match. This argument overlooks the fact that even a small deterrent force can be extremely effective since the utility of the nuclear weapon lies more in the manner in which it is politically manipulated rather than in its actual use. And even in the remote, extremely unimaginable situation of its actual use, a simple fission device delivered on target could serve the purpose. New Delhi could hardly be happier to have killed with a more powerful thermonuclear weapon than with an 'ordinary' fission one! At no time must India lose sight of the fact that nuclear weapons are more political tools than otherwise.

Therefore, with or without international cooperation, India's fissile material production and stockpiling trajectory is unlikely to be very different. There is nothing to indicate that even in the absence of safeguards, India would use all its PHWRs for strategic purposes. Just as there is nothing to suggest that with the application of safeguards, India would have to stop doing something that it was doing in the past. Of course, the deal would impact India's nuclear programme by way of the logistics and financial costs of separation, but it is unlikely to either favourably or unfavourably impact the existing fissile material stockpiles or the future stockpiling capabilities. These, instead, could be more affected by the conclusion of an FMCT, at some date in the future because that would constrain the right available to India under the bilateral cooperation agreement to build more military designated reactors. This issue too has been addressed in the previous chapter.

Conclusion

Given the historical path followed by India's nuclear programme, the energy and strategy dimensions are closely interlinked. While the military programme has gained from the civilian developments, the power programme has often found itself a prisoner of sanctions and denials imposed because of India's pursuit of the military nuclear programme.

The present circumstances offer a historic opportunity to delink the two programmes – through a separation of facilities. This would set the power programme free to pursue rapid additionalities. Of course, this would mean that the military programme would have to build some of its own facilities that it was piggybacking upon earlier. However, this step would prove to be beneficial for both in the long run. By disassociating itself from the military dimension, the power programme will be free to access international nuclear commerce and, thus, enhance the share of nuclear energy in the total energy production of the country. This is a critical requirement given that of all the challenges facing an emerging India, two could particularly derail the process of its rise, and they are interconnected. The first is widespread social disharmony as a result of skewed economic development and unmet aspirations and expectations of a growing population. The resulting discontentment coupled with growth of sub-nationalism, opportunistically exploited by domestic vested interests or external adversaries could severely strain the country's socio-politico-economic fabric. Hence, the importance of social and human development along with economic growth in order to ensure societal peace and national security. The second challenge arises from the sheer shortage of power, especially electricity, to drive the economic processes and developmental endeavours. It is widely accepted that electricity has a direct connection with level of development and per capita energy consumption is a parameter for calculating the human development index. Therefore, it is hardly surprising that energy and the ability of a nation to access it from reliable, secure, sustainable and safe sources tops national priorities everywhere. This is even more important for a nation like India that expects a phenomenal growth in its energy demand

In the long run, therefore, the strategic prudence of placing some of the nuclear reactors under safeguards and fire-walling them from the strategic programme, rather than subjecting the entire country to the risks of energy vulnerability and possible economic stagnation will become evident. The nuclear key would help open the door to other high technologies that are required for a fast maturing economy in need of constant technological upgradation but constrained by a continually expanding regime of technology denials. As the dangers of WMD proliferation become more palpable, and nuclear disarmament remains a distant dream, supplier nations will try to control technologies even further. India has done well to seize the opportunity in 2008 to break out of the technology denial stranglehold that had limited its opportunities since 1974.

10

The Shadow Dance

Nuclear Deterrence and Conventional War

The essence of power is to know the limits upon power. The linkage between nuclear deterrence and conventional war rests upon this crucial understanding. While the relationship between the nuclear weapons of two nations and the establishment of credible nuclear deterrence is fairly well understood owing to the Cold War experience between the superpowers, the practice of nuclear deterrence with respect to conventional war is a subject that has been far less studied. The US and the USSR deterred each other at the nuclear level and the conventional wars between them were fought by their proxies, not by them directly. These were wars between nations that were themselves not nuclear and of the kinds that never really threatened the territorial security or survival of the superpowers. If that was threatened, as in the case of the Cuban missile crisis in 1962, then it was expected that the breakout of any conventional hostility would rapidly escalate to the nuclear level. Given this presumption, the focus then shifted to the very conduct of nuclear war itself and to issues specific to such war-fighting, e.g. the use of tactical nuclear weapons, efficacy of first strike or counter-strike doctrines, calculation of number of nuclear weapons required to prevail and claim victory in a nuclear war, etc. Caught in this maze of nuclear war-fighting, little attention was paid to the conduct of conventional war in the presence of nuclear weapons – a situation where conventional hostilities did not move up to the nuclear level.

India's practice of nuclear deterrence, however, deals with a unique predicament where its nuclear weapons are certainly meant to deter the adversary's nuclear use or blackmail. But, at the same time, especially with Pakistan, they are tasked to function in a fragile situation where Pakistan's nuclear weapons seek to neutralise India's conventional superiority even as it engages in sub-conventional conflict with impunity. Therefore, the interlinkages among sub-conventional, conventional and nuclear war are far more complex in the Indian case. In fact, India's nuclear strategy has to grapple with the challenge of building the credibility of its nuclear deterrence in such a way as to counter the adversary's attempt to blur the lines between conventional and nuclear war. While he threatens to lower the bar for breakdown of nuclear deterrence in order to deter conventional attack, India must not only raise the nuclear threshold, but also devise adequate conventional responses that can be safely executed in the situation of nuclear overhang. Of course, this is easier said than done. As long as nuclear weapons exist, so does the potential danger of escalation to nuclear use. But, the true test of India's nuclear strategy lies as much in establishing mutual nuclear deterrence, as in tackling sub-conventional warfare with conventional tactics that are conscious of the presence of nuclear weapons, but can operate without bringing them into the calculations.

Is this possible? What impact do nuclear weapons have on the conduct of conventional warfare? What are the relevant lessons that can be discerned from the experiences of Kargil and Operation

Parakram? What kind of an option does 'limited war' offer? How best can India use this option to nullify the advantage of nuclear weapons that Pakistan seeks? These are some of the questions that this chapter grapples with.

Advent of the 'Absolute Weapon'

The dropping of the two atomic bombs, Little Boy and Fat Man, on Hiroshima and Nagasaki starkly illustrated the destructive potential of the new weapons. Even more significantly, the two bombs showcased the power they brought to the possessor to influence the behaviour of the adversary by implicitly or/and explicitly threatening the use of such horrific destruction. In one instant, therefore, nuclear weapons altered the criteria for measuring national power and the dynamics of conflict interaction in international politics.

Before the advent of nuclear weapons, states could risk war or even engage in it for a stake they felt was high enough to justify the action. After all, according to the Clausewitzian concept, war was only an extension of politics by other means. So when politics was unable to achieve a desired objective, war was the tool to be employed. But nuclear weapons completely changed this equation. Their high destructive potential compelled one to pause and question the value of the interest at stake and the potential cost of the war. In fact, war in a situation where nuclear weapons were present with both sides, created the inevitability of mutual assured destruction (MAD), thereby nullifying its use as politics by other means. Consequently, the purpose of the new weapon, for the first time in military history, became to avoid its use instead of crafting strategies to incorporate it into military operations. Bernard Brodie captured the essence of this thinking soon after 1945 in his seminal work *The Absolute Weapon*. He argued that the purpose of the nuclear weapon, the *absolute weapon* as he described it, was to prevent wars, not fight with them. This basic premise has remained unchanged since then. True to his original conclusion, over the last six decades, countries possessing nuclear weapons have used them for deterrence against other possessors of these WMD, or for compellence against non-possessors, but not for war-fighting.

Nuclear deterrence may be described as a strategy that stops an adversary from indulging in any nuclear use by instilling the fear that the cost of the action would entail assured damage of a kind that would be much more than the gain sought to be made. Of course, it is a different matter that in order to prevent breakdown of nuclear deterrence, countries project the image that they are prepared to handle just such a situation, which means being ready to use the weapon. As per the strange logic of nuclear deterrence, the equation is simple: better planning for deterrence failure leads to higher credibility of deterrence, which in turn reduces the possibility of deterrence failure. Driven by this paradoxical reasoning, the superpowers, during the Cold War, consistently expanded and upgraded their nuclear arsenals and altered doctrines and strategies to accommodate new capabilities. While deterrence was the primary focus of their foreign policy and bilateral interactions, nuclear capabilities were nevertheless always shown as being ready to be pressed into military use, if required. Elaborate standard operating procedures were adopted by both nations to indicate an almost automatic launch of nuclear weapons once the decision had been made.

However, the fact that no actual use of nuclear weapons[1] ever occured is taken as the success of nuclear deterrence. It is widely

1. There are several instances, however, of the use or threat of use of nuclear weapon for blackmail and coercion. Until 1996, a list of 47 identifiable such incidents had been compiled by Jasjit Singh, "India: The Sixth State with Nuclear Weapons," *Asian Strategic Review 1997-98* (New Delhi: IDSA,1998),pp.12-13.

assumed that the presence of these WMD and the concomitant fear of MAD kept the superpowers from engaging in any direct confrontation. Of course, the nuclear stability at the level of the two rival superpowers contrasted with a more volatile situation at the lower levels of their allies. This came to be known as the "stability-instability paradox" and was used to explain the proxy wars that happened between the two ideological blocs. But, the risk of escalation and cost of miscalculation cast a constraining influence on state behaviour and kept them away from any direct confrontation.[2]

The only instance of two nuclear countries fighting a direct war with one another, prior to 1999, was the Ussuri river conflict between China and Soviet Union in 1969. Triggered off by an ambush of Soviet troops by Chinese forces at Zhenbao island along the Sino-Soviet border, it led to the killing of 31 Russian soldiers. The Chinese action which, according to their accounts, was a response to Soviet provocations along the border from mid-1960s onwards, was taken in the belief that "the border clash was a *controllable military conflict*" that would serve the "larger domestic political purpose of mobilizing the Chinese people for further revolution."[3] However, the Chinese act was not perceived as a limited attack in Moscow, and the Soviet leaders did consider a number of military options, including a disarming strike on China's nuclear arsenal. The tension lasted several months, during which the Soviet Union even probed the US on its response if the Soviets were to attack China's nuclear facilities. The US remained ambiguous, and both

2. This is best illustrated in the experience of the Cuban missile crisis of 1962 when despite their 'macho' nuclear doctrines, both the US and USSR exercised extreme caution and accepted compromises to keep a direct confrontation at bay.

3. David Ochmanek and Lowell H. Schwartz, *The Challenge of Nuclear-Armed Regional Adversaries* (Santa Monica, CA: RAND Corporation, 2008), pp. 33-34. Emphasis added to highlight the perception.

China and Russia refrained from taking very provocative actions. The crisis finally wound down after the Chinese Premier Zhou Enlai categorically stated to the Soviet Premier that China had no aggressive intentions and that its nuclear weapons posed no threat to the USSR. As the crisis subsided, it revealed the overarching influence of nuclear weapons on two adversaries when both possess nuclear weapons.

Conventional War in the Presence of Nuclear Weapons

In 1991, Martin van Creveld, a well known analyst on war, wrote, "From Central Europe to Kashmir, and from the Middle East to Korea, nuclear weapons are making it impossible for large sovereign territorial units, or states *to fight each other in earnest without running the risk of mutual suicide*."[4] Of course, a number of factors other than nuclear weapons are also responsible for changing the nature of warfare from total to limited wars in contemporary times.[5] However, the most important limitation on war in situations where both adversaries possess nuclear weapons is cast by the presence and impact of these weapons. While wars may still have to be fought, the shadow of nuclear weapons, nevertheless, imposes constraints on the range of military options and the nature of coercive use of force that nations can indulge in. It demands greater caution so as to avoid the potential cost of miscalculation. Leaders of nuclear-armed nations must be constantly aware of the risks involved, especially in a show of force, and are required to walk a tightrope in case of a crisis. On the one hand, the desire to win the dispute requires a demonstration of resolve and a willingness to fight. But simultaneously, the fear

4. Marin van Creveld, *On Future War* (London: Brassey's, 1991), p. 194. Emphasis added.
5. For a detailed analysis of the many factors that are leading the trend to limited wars, see Jasjit Singh, "The Dynamics of Limited War," *Strategic Analysis*, vol. 24, no. 7, October 2000.

of nuclear war demands caution and restraint in the use of force. Unrestrained coercive manoeuvring by either or both sides could end up committing them to a war neither wanted in the first place. And the risks of such unplanned and uncontrollable escalation in the presence of nuclear weapons could be catastrophic for both.

Therefore, the conduct of war in the presence of nuclear weapons has to follow a different set of rules. At one level, in fact, nuclear weapons mean the end of classical conventional war of the kind envisaging acquisition of large swathes of territory, or a *blitzkrieg* effort to cause high attrition. Any such measure is certain to breach the threshold of the adversary's levels of tolerance, especially one with weaker conventional capabilities. This could increase his dependence on the resort to use of nuclear weapons, thereby leading to deterrence breakdown. If deterrence is to be maintained, then the war needs to be fought differently, in a manner whereby the risk of escalation to the nuclear level is minimised because the targets are chosen so as not to threaten the survival of the state or its critical elements.

This obviously has an impact on how victory and defeat can be defined in such a situation. In an all out conventional war, the difference between the victor and the vanquished is clearly evident based on the estimation of which side has suffered greater losses and damage. But, in limited wars, in which the level of destruction has to be carefully calibrated and imposed, this distinction is blurred. In fact, as explained by Jasjit Singh, "As war starts to move down the intensity spectrum, victory and defeat shift more into political and psychological dimensions."[6]

An illustration of this is manifest in the experience of India and

6. Ibid. It is also worth noting here that in modern warfare, the media plays a very important role in declaring who the winner is. It shapes perceptions and public mood that could sometimes be quite different from the reality on the ground. The 2007 Israel-Lebanon War presents a good illustration of this point.

Pakistan during the Kargil crisis in 1999 and Operation Parakram in 2001-02. Coming soon after the overt demonstration of nuclear capability by India and Pakistan in May 1998, the start of these crises was greeted by strategic analysts everywhere with great pessimism and they were quick to dub them as the realisation of their worst fears when regional powers acquire nuclear weapons. Given the prevailing regional reality where nuclear-armed adversaries shared unsettled territorial disputes, a relationship marked by mutual hostility and distrust, routine border skirmishes and terrorist violence as a result of the Pakistan-fomented proxy war in India, the apprehensions of the international community of conventional war quickly escalating to the nuclear level were perfectly understandable. However, the manner in which Kargil and Operation Parakram unfolded holds several lessons for Pakistan, India and the larger international community on how use of force is possible even in the shadow of nuclear weapons.

The Experiences of Kargil and Operation Parakram

In May 1999, barely one year after going nuclear, Pakistan infiltrated approximately 5,000 soldiers in the guise of mujahideen across the Line of Control (LoC) separating the Indian and Pakistani controlled regions of Kashmir. Their mission was to seize strategic pieces of territory and then compel the Indian government to negotiate the future status of Kashmir. Pakistani leaders believed that their recently demonstrated nuclear capability would deter India from using its conventional military superiority against Pakistan.

This confidence was derived from the predominant understanding of the Pakistani nuclear doctrine which encapsulates a more offensive form of deterrence that seeks to change the *status quo* by holding out the threat of nuclear blackmail while deterring an Indian conventional attack. As a senior Pakistani official maintained, "The Indians cannot afford to extend the war to other areas in Kashmir, leave aside launching an attack across international boundaries" because of the "risk of nuclear conflagaration."[7] By suggesting this linkage time and again, the Pakistan Army assumes that it could raise the military and economic costs for India without endangering its own security to the risk of retaliation against the proxy war. By making nuclear threats, it seeks to manipulate risks to its advantage even if following through on them would be nothing short of suicidal for itself. Whether this would really happen or not, Pakistan banks on uncertainty or irrationality about its actions for deterring India.[8]

Subscribing to this logic, Islamabad assumed during its Kargil operations that India would find its military options checkmated by the presence of a nuclear overhang and would be compelled to negotiate. The achievement of a rough nuclear parity was considered enough to offset asymmetry in conventional capabilities. Also, given that ever since May 1998, an anxious US had consistently described South Asia as a dangerous nuclear flashpoint, Pakistan hoped that Washington would not hesitate to intervene to resolve the crisis and even formalise a new status quo in Kashmir, which could then be claimed as victory for Pakistan. This internationalisation, it was naturally believed, would work in Pakistan's favour.

However, that is not how the saga unfolded. India's response, hesitant in the first two weeks whilst the military still believed that the action in the area was by unusually well armed irregulars and terrorists, became far more considered and calibrated as soon as greater clarity about the ground situation emerged. The Army and Air Force undertook coordinated actions with a clear

---

7. Quoted in Zahid Hussain, "On the Brink," *Newsline*, June 1999, pp.24-25.

8 Interestingly, in his book *In the Line of Fire*, Gen Musharaff has alluded to his commando personality that prodded him to be a risk taker.

objective of dislodging the Pakistani forces without foreclosing any future options. But, the political leadership imposed on them the constraint of confining their military operations to the Indian side of the LoC. While this imposition came with its challenges, two factors worked in India's favour. Firstly, having first claimed that the intruders were not really its own troops but independent mujahideen, Islamabad found itself constrained to provide its troops with any reinforcements to fight the Indian military offensive that came in the form of artillery barrages as well as air attacks. Secondly, once the international community knew of the subterfuge, Pakistan was seen as the aggressor and pressured, even by its traditional friends, Washington and Beijing, to withdraw from the occupied heights. In fact, the US, which had shied away from assigning blame to a particular party for having initiated a crisis in previous wars between India and Pakistan, squarely condemned Pakistan's incursions and intentions across the LoC and refused to mediate. China too counselled Prime Minister Sharif, during his visit to Beijing, that Pakistan should abandon its plans. A more sympathetic response from its traditional allies might have emboldened Pakistan to carry on. But faced with escalating casualties, economic losses due to sanctions and diplomatic isolation, Prime Minister Nawaz Sharif was compelled to reconsider continuing military operations, even in the face of advice to the contrary by his own military commanders.[9]

---

9. In fact, there has been a war of words between Pakistan's political and military leadership of the time with each blaming the other for the Kargil debacle. Gen Musharraf has never tired of recounting the "political mishandling" of the situation and attempts to "spin the events disingenuously". See Pervez Musharraf, *In the Line of Fire: A Memoir* (New York: Free Press, 2006), pp. 95-98. On the other hand, then Prime Minister Nawaz Sharif has maintained that he was never briefed on the military operations in Kargil. Meanwhile, Benazir Bhutto has stated that the army had shared the plan with her when she was Prime Minister, but that she had vetoed it even though she had sensed that the Army Chief was taken in with the "brilliance of military

The war highlighted the politico-diplomatic dimensions of modern wars, especially where nuclear weapons are concerned. Even as the Indian military moved on the ground to oust the infiltrators, attempts were simultaneously mounted to diplomatically isolate Pakistan and expose its offensive designs to alter the status of the LoC, a line whose sanctity had long been upheld as the very basis of a negotiated settlement for the disputed territories in the region. In fact, Kargil, for the first time, made India reap the benefit of "internationalisation" of the Kashmir issue, a ploy often used by Pakistan and desisted by India. In this instance, the US intervention sought by Pakistan turned into India's advantage for two reasons: one, as the Indian Army routed back Pakistani forces and moved closer to the LoC, the risk of extension of war across the LoC control increased. With this also rose the fear of military escalation. Having already limited the Indian Air Force (IAF) to operating in difficult situation on its own side of the border, New Delhi was willing to accept US intervention to control Pakistan; second, flowing from the show of Indian maturity and self-control in handling a provocative situation, the US' perception of India and its nuclear status underwent a change. This enabled the blossoming of a strategic relationship that has been far more understanding of the Indian security environment and compulsions.

Kargil ended in July 1999, roughly two months after it had started, as an ill thought out misadventure by Pakistan. However, it was useful to the extent that it illustrated (to those on both sides of the border who are willing to rationally assess it) the limits that nuclear weapons cast on actions of nations. Pakistan realised that the acquisition of nuclear weapons had not provided it with

---

strategy". See Samina Ahmed, "Nuclear Weapons and the Kargil Crisis," in Lowell Dittmer, ed., *South Asia's Nuclear Security Dilemma: India, Pakistan and China* (London: East Gate, 2005), p. 143.

a *carte blanche* on disruptive actions across the border. Rather, the presence of nuclear weapons placed clear limits on how far it could, or should go so as not to breach the limits of Indian tolerance. The cuts, therefore, that it inflicts on India must not be deep enough to touch a raw nerve and make it squeal in pain. But they could be several and skin deep to cause a constant ache and needing constant attention.

Meanwhile, the Kargil episode made India realise the constraints that the presence of the adversary's nuclear weapons cast on its own exercise of military options. Despite the widely expressed opinion to strike against Pakistan once the identity of the mujahideen as regular Pakistani soldiers was established beyond doubt, the political leaders directed the military to undertake operations in such a way that the threat of escalation was minimised. Therefore, in an unprecedented gesture, the use of air power was limited to the Indian side of the LoC in order to oust the illegal occupants of the heights. No strikes were authorised across the border, not even at the terrorist infrastructure known to exist in Pakistan Occupied Kashmir (POK). This was in stark contrast to the Indian response to Pakistan's Operation Gibraltar in 1965. Even then, Gen Ayub Khan's military regime had sent Pakistani regular forces disguised as Kashmiri dissidents into the region. This operation was premised on the belief that India would not have the stomach to spread the conflict beyond the disputed territory and, thus, enable Pakistan to succeed with its revisionist plans of occupying some chunks of territory. But, at the time, India, despite its weak military position, so soon after the 1962 defeat at the hands of China, had not hesitated to extend the conflict beyond the international boundary.

The case with Kargil was just the opposite since despite its position of conventional superiority, India exhibited the ability to wage a war with self-imposed limits. This proved to be as much a revelation to Pakistan as to the larger international community that had described this very region as the most dangerous flashpoint. The sense of responsibility and maturity in action displayed by India helped to shape a range of perceptions across several capitals. The NPT ordained nuclear weapon states realised that nuclear weapons in the possession of regional powers were no more or less dangerous than when in their own arsenals. Islamabad and New Delhi, meanwhile, realised the strengths and limitations of nuclear weapons.

Some of these lessons were again put to the test in December 2001 after the Pakistan-supported terrorist attacks on the Indian Parliament. It is a striking coincidence that this incident took place three months after 9/11 when a hijacked airliner was believed to be headed for a collision with the Capitol Hill, the seat of the American Parliament. While the aircraft failed to reach the target, the terrorists in New Delhi partially succeeded in breaching the security of the Indian Parliament and managed to reach one of its gates before being shot dead. However, a strike against a potent symbol of Indian democracy, and at a time when Parliament was in session, the incident ignited aggressive sentiments from the Indian polity, public and military. A significant section of the political voices, military advice and editorial opinion clearly argued in favour of targeting terrorist training camps in POK. In a show of resolve, the political leadership did authorise the mobilisation of military forces on the international border and the LoC in Kashmir. The move elicited a reciprocal response from across the border. The international community watched with concern as Operation Parakram unfolded. At one time, some governments even issued travel advisories to their nationals visiting or residing in this part of the world to indicate the seriousness of the situation between two nuclear-armed nations with fully mobilised militaries on high alert. However, despite the high level of military mobilisation, India and Pakistan did not actually go to war. While this is attributed to a range of reasons, including external influence, diplomatic parleys, economic constraints, etc., the presence of nuclear weapons is not something that can be overlooked.

In this instance, India practised compellence or what has been described as "coercive diplomacy"[10] that involved a combination of diplomatic and military pressure. The basic aim was to force the Pakistan government to accede to India's demands of halting all support to cross-border terrorism and to take action against terrorist outfits known to be operating from its soil or else be prepared for military action. However, Jaswant Singh, then Foreign Minister, has listed three aims for the mobilisation[11]:

(a) To defeat the cross-border infiltration/terrorism without conflict. To this end, the government tried to contain Pakistan diplomatically.
(b) To contain the national mood of 'teach Pakistan a lesson'. This required managing the country's sense of outrage and desire for revenge and retaliation by not necessarily waging a war but providing a sense of achievement to the country by diplomatically defeating the enemy.
(c) To destroy and degrade Pakistan's war-fighting capabilities in event of war. In readiness for such an eventuality, the military was placed on a state of high alert. However, the challenge lay in getting the armed forces leadership to recognise the value of restraint as a strategic asset.

As is often the case in any war, these aims appear far more clearly articulated in retrospect than they came out at the time of the crisis. At that time, the Indian policy seemed a bit rudderless,

10. Jaswant Singh, *A Call to Honour* (New Delhi: Rupa Publications, 2006), p. 266.
11. Ibid., p. 268.

especially once Islamabad denied, *ad nauseam*, its involvement in any such activities and reciprocated with its own military preparations. At this juncture, the presence of nuclear weapons did compel India to reconsider and reevaluate its military options because neither could it ratchet up the crisis given the possibility of the conflict escalating to the nuclear level and leading to destruction far higher than anticipated, and nor could it wind down the mobilised war machinery without having achieved the political objectives first articulated. Some of these were claimed to be met on January 12, 2002, when President Musharraf in an address to the nation announced that Pakistan would no longer be used as a base for terrorism of any kind, and that it would ban Lashkar-e-Tayyeba and Jaish-e-Mohammed implicated in the attack on Parliament. However, the Indian military remained mobilised and the Indian resolve to take action against Pakistan strengthened after the Kaluchak terrorist massacre in May 2002 when families of troops mobilised at the border were killed. Help from the US at this point to coerce President Musharraf to make a public statement decrying terrorism and banning the outfits came in useful in defusing the tension. After Pakistan's commitment made in a public announcement in early June 2002 to halt its support for cross-border terrorism, the war machinery was wound down.

In both cases, Kargil and the attack on the Indian Parliament, Islamabad was emboldened to take the step in the belief that the hands of India's military would be tied, preventing it from taking any decisive military action against Pakistan. Simultaneously, Pakistan played the nuclear card at the international level, leaning heavily on the international community to restrain India if it wanted to avoid nuclear use in the region. Pakistan's Ambassador to the UN made it clear in May 2002 that Pakistan would not "expend our limited resources on building up a conventional defence which will completely debilitate our development… We have to rely on our means to deter Indian aggression. We have that means and we will not neutralize it by any doctrine of no first use."[12]

Indeed, India did feel the weight of nuclear weapons on its possible courses of action. But it also discovered that there was scope for retaliatory action that had to be intelligently discovered and exploited and astutely meshed with politico-diplomatic measures best suited to the prevailing international political environment. This, if effectively employed, brought the possibility of achieving important political gains without actual resort to military action, or *use of force without war*. Pakistan failed to understand this basic feature of modern warfare during Kargil and Operation Parakram. As stated by an American strategic analyst, "The loss of its vital US ally and the non-appearance of Chinese support suggest a poor appreciation for alliance considerations and international reactions to the attack."[13] The same experience has been repeated for India and Pakistan after the 26/11 Mumbai terrorist attacks on two prominent hotels and a Jewish centre.

India needs to understand the role of external players in a bilateral relationship between two nuclear-armed neighbours. While such interference and the possibility of "internationalisation" of the Kashmir dispute were considered anathema to the Indian administrations in the past, the experience of the three crisis situations post 1998 have shown the benefits of US mediation. This is not to suggest that India should depend on the US or other states to deal with its security concerns vis-a-vis Pakistan, but to highlight that in the Indo-Pakistan relationship, especially in the presence of nuclear weapons, the role of external powers would be greater given that their apprehensions of nuclear

12. As quoted by Timothy D. Hoyt, "Strategic Myopia: Pakistan's Nuclear Doctrine and Crisis Stability in South Asia", in Dittmer, ed., n. 9, p. 120.

13 Ibid., p. 130.

use are more. India has less to fear from this influence in the future since a mature handling of the situations by India has helped create the distinction in the American mind on the behaviour of the two nations. Therefore, the US that earlier never tired of straitjacketing its nuclear policies into the same mould for India and Pakistan is today willing to treat them differently. An adept use of growing Indian influence with the US could certainly help to keep Pakistan from playing the game of nuclear brinkmanship.

The Conduct of Limited War

As explained in the previous section, the nature of warfare changes with the entry of nuclear weapons into the calculations, and, hence, the concepts, doctrines and capabilities must also keep apace. In fact, the challenges are further complicated since the presence of nuclear weapons considerably raises the threshold of provocation. So, Pakistan, for instance, is able to exploit far greater space at the lower level of sub-conventional conflict to indulge in acts seeking to destabilise India. It derives immunity against conventional war by raising the risk of escalation to the nuclear level. This is a strategy that Pakistan has crafted to perfection. The challenge for India, therefore, lies in nullifying the advantage that the adversary seeks to exploit from the linkage between nuclear deterrence and conventional war.

Limited war offers one possible response to the situation at a minimum risk of nuclear escalation. As the very term indicates, limited war means a restrained, calibrated use of force instead of an all out employment of military capabilities. Normally, militaries do not welcome constraints on the use of their resources. For them, the achievement of victory in war is the final and singular objective and their military might is meant to be effectively used as a tool in the pursuit of this goal. However, if an all out war was to be fought with nuclear weapons, then victory at the cost of losing a nation would appear self-defeating, if not downright foolish. Hence, the significance of limited war. Obviously, this would involve a revision of conventional warfare tactics that hold the risk of breaching the adversary's nuclear threshold to a version where short, sharp wars limited in time, scope and intent are undertaken. Therefore, military strikes would either need to be restricted in depth into enemy territory and spread in geographical expanse, or limited in scope to carry out deeper, narrow thrusts into adversary territory in order to remain well away from the expressed 'red lines' of the nuclear threshold.

In the case of Pakistan, with which the chances of deterrence breakdown are deemed to be higher than with China[14], red lines of some sort were spelt out by Gen Khalid Kidwai of the Strategic Plans Division, the organisation in charge of Pakistan's strategic assets and policy in 2001. These were: loss of large parts of territory (space threshold); destruction of large part of land or air forces (military threshold); economic strangulation (economic threshold); and political destabilisation or large-scale internal subversion (domestic destabilisation threshold). Evidently, the broad areas covered by this articulation are clearly meant to indicate a very low nuclear threshold for Pakistan's nuclear use. However, it is imperative that the Indian national security establishment draws up its own estimation of the credibility of these threats. For instance, on the space threshold, it can be safely determined that even a deep penetration by India into the

14. Even though China is considered the larger potential threat for Indian security owing to the possibility of clash of interests for resources and spheres of influence in the coming decades, it is with Pakistan that the dangers of deterrence breakdown are more pronounced owing to its pursuit of asymmetric warfare with India and a more aggressive offensive defence military strategy. Meanwhile, given the doctrinal similarities of no first use and the acceptance of nuclear weapons as a political tool of deterrence rather than war-fighting, nuclear deterrence between India and China appears far more stable.

barren desert area of Pakistan is unlikely to breach its nuclear threshold, though a similar depth in the populated areas of Punjab would be viewed differently. Similarly, air strikes against terrorist infrastructure and assets in POK are likely to evoke less response than targeting military assets elsewhere in Pakistan. Such assessments are necessary in the case of every threshold if India is not to be self deterred from undertaking action against the provocative conduct of sub-conventional war by Pakistan. The possible military response, meanwhile, would have to be in keeping with the concept of limited war along the following parameters.

One, there is an urgent need to reconsider the existing model of troop mobilisation. As was learnt from the experience of Operation Parakram, full mobilisation of that kind in the presence of a nuclear-armed adversary is hardly an effective proposition, besides being dangerous. Moreover, display of such military preparedness in the absence of political resolve also sends wrong signals to the adversary, thereby degrading the deterrence at even the conventional level. Therefore, the Indian military has to look for options that allow it to undertake short, shallow strikes/thrusts into adversary territory in areas that would make a difference.

Speed of action is also of the essence in such operations. The recently articulated (April 2004) Cold Start doctrine of the Indian Army proposes some fresh military options to address the challenge. Drawing upon the Warsaw Pact strategy of "swift advance [with] the ability to attack from a standing start,"[15] the basic idea of Cold Start is to have a ready, mobile lethal force of eight integrated battle groups with integral logistics capable of mounting an offensive with air cover. The idea behind this plan is to avoid a long preparation time and to advance quickly into

15. Bharat Karnad, *India's Nuclear Policy* (New Delhi: Pentagon Press, 2008), p. 115.

enemy territory for shallow penetration attacks. This would place the onus of escalation of hostilities on the adversary while retaining the initiative with oneself. Of course, the plans would require meticulous conceptualisation and preparation for translating them from good theory to effective practice.

In fact, even more than the use of the army and its new doctrine, it may be prudent to consider using those arms of the military that offer the maximum possibility of highly calibrated escalation and the ability to de-escalate in these operations. Therefore, use of such instruments as Special Forces (especially raised and trained for the purpose) or air power or even maritime power with the requisite capabilities would be preferred options. Such employment of force could be best conducted with maritime or air power because they enjoy, in varying measure, the advantage of flexibility of employment, as well as better control over military engagement and, hence, over escalation. Air power provides the greatest benefits in this regard while land forces have little advantage in terms of escalation control. Once engaged in combat, the army cannot be disengaged unless one or the other side either concedes defeat or a ceasefire is agreed upon. Menawhile, air power helps show resolve, and, at the same time, offers the flexibility of disengagement, thereby making it possible to control escalation. Therefore, as a second measure, adequate attention must be given to understanding the advantages and limitations of every Service in different scenarios. Such issues need to be adequately considered and deliberated upon in peacetime in order to meet the requirements during a period of crisis.

Thirdly, India must consistently build up its conventional capability in order to keep the nuclear threshold as high as possible. The thrust areas for modernisation must include reconnaissance, surveillance, and intelligence capabilities that can enable precise target acquisition and attack, as well as air attack capabilities with precision guidance in order to minimise risk of collateral damage. This would not only reduce the chances of escalation but also garner greater acceptance from the major powers that could then be counted upon to bear down upon the adversary to see reason and temper its responses accordingly.

Fourthly, and most importantly, it is imperative that India continuously works at enhancing the credibility of its nuclear deterrence. This demands moving towards a survivable second strike capability. Given its no first use posture, India has committed itself to a retaliation only policy and, hence, is required to concentrate on increasing the survivability of not just the warhead or the delivery vehicle but also the entire command and control structure, communication networks and, above all, the survival of the political will to retaliate. Survival of the weapon would mean little in case the political leadership is not adequately prepared to understand the demands of nuclear deterrence. Also, it must be appreciated that political will in a democracy depends a great deal on the perceived legitimacy of action. National will arises from, and can be built by articulating and encouraging a clear understanding of national interests and policy options to pursue them. These are challenging issues and must be accorded the attention they deserve.

Fifthly, along with military preparedness, it is critical to showcase firm political resolve and determination to achieve objectives that themselves need to be clearly articulated for the adversary and for the domestic audience. Clarity on the objectives of the military exercise would provide an indication to the adversary that the goals of the operation are strictly limited and, hence, there is no intention to breach or even reach near his stated or perceived red lines. Also, the clarity would enable better management of domestic expectations thereby garnering legitimacy and support for the actions.

In conclusion, it may be stated that while nuclear deterrence imposes restraints on conventional war, it is not possible for all wars to be obviated in their presence. In difficult regional circumstances such as the kind that India finds itself in, the possibility of conventional wars exists. Theoretically, therefore, there is also the possibility of escalation of a conventional conflict into an unwanted nuclear exchange. However, if India was to be self-deterred by this thought, it would mean complete erosion of both its conventional and nuclear deterrence capabilities. In order to deter war, India must maintain and project a high level of conventional capability that would be intelligently applied in a calibrated manner to keep the use of coercive force well below the assumed red lines of the adversary. In order to deter nuclear war, India must illustrate its ability to handle deterrence breakdown and retaliate against the adversary with enough capability and resolve to inflict damage that would impose a cost far beyond the value of the stake that made the first use of the nuclear weapon against India thinkable.

11

Finding the Way

From Deterrence to Disarmament

Having discussed some of the critical issues on how India should build its nuclear security in the presence of nuclear weapons with its adversaries, this chapter attempts to explain why the ultimate, final nuclear security for India lies in the universal abolition of nuclear weapons and how India could continue to meaningfully contribute towards this endeavour. It is an unfortunate reality that as long as nuclear weapons exist with those whom India perceives as its security threats, it too will have to maintain its own nuclear weapons in order to establish credible nuclear deterrence. This is a compulsion that nuclear weapons impose since there is no other weapon that can deter nuclear weapons. In order to avoid nuclear weapons being used against oneself, one is compelled to acquire the same capability. Therefore, the actuality of the day has to be met realistically. This, however, need not stop India from aspiring for an ideal that it believes holds the promise of better security for itself. Even if that goal appears unfeasible at the moment, if it is desirable, it is certainly worth pursuing in whichever manner is available within the constraints of the existing realities.

Therefore, there is no inherent contradiction in India's stand on nuclear disarmament, as sometimes alleged. The country has consistently argued that there is greater security – national and international – to be found in a nuclear weapon free world (NWFW). And, so in all its interventions on the subject, at every international forum, New Delhi has argued for, and made innovative suggestions for, the realisation of just such a world. Pre-1947[1] and after independence[2], pre-1967 and after the conclusion of the NPT[3], pre-1974 and after its peaceful nuclear explosion[4], pre-1995 and after the indefinite and unconditional extension of the NPT, pre-1998 and after the overt demonstration of its nuclear weapons capability, every Indian government has maintained a desire for nuclear disarmament. This was well articulated by Prime Minister Vajpayee at the time of India's five tests when he said,

> All the governments in New Delhi since Independence have steadfastly maintained that our national security as well as global security is guaranteed only in a nuclear-weapon-free-world. India backed up this conviction with a large number of

1. Jawaharlal Nehru had said in 1940, "Both because our adherence to the principle of non-violence and from practical considerations arising from our understanding of world events, we believe complete disarmament of all nations should be aimed at, as in fact an urgent necessity, if the world is not to be reduced to barbarism," A confidential note written in Wardha on August 25, 1940, and as quoted in *India and Disarmament: An Anthology* (New Delhi: Ministry of External Affairs, 1988), p. 11.
2. In 1953, speaking before the First Committee of the UNGA, then Foreign Minister V.K. Krishna Menon said, "There is only one thing to do with atomic weapons and that is to do away with them."
3. In 1964, India took the initiative to place the item "Non-proliferation of Nuclear Weapons" on the UN agenda and thereby sowed the first seed of the NPT.
4. At the UNGA's First Special Session on Disarmament in June 1978, then Prime Minister Morarji Desai called for outlawing the use of nuclear technology for military purposes. A decade later, Rajiv Gandhi presented a comprehensive Action Plan for Ushering in a Nuclear Weapon Free and Non Violent World Order to the Third Special Session on Disarmament.

> initiatives in the United Nations and other forums... Sadly our initiatives were disregarded by the NWS...Yet, India waited 50 years before responding.

Before India conducted the tests in May 1998, there were many who dismissed India's initiatives towards elimination of nuclear weapons as a case of sour grapes. It was said that since India did not possess nuclear weapons, it was canvassing for their elimination so that none would have the advantage. After 1998, there are many, especially within the country, who argue that India must now give up its moralist stance on nuclear disarmament and concentrate instead on building an effective and reliable deterrent. This chapter, however, argues that for India, its struggle for universal nuclear disarmament goes beyond the rationale spelt out by these two arguments. The grapes remain equally sour, or the nuclear weapons as horrendous, even when India has them. Meanwhile, India's position on universal nuclear disarmament is not only a morally correct one (and the realists almost make one feel apologetic about it), but it is also simultaneously rooted in the national security interests of the country.

In the wake of vertical and horizontal proliferation in the neighbourhood, India was left with little choice but to go down the nuclear path itself since the only effective deterrent for nuclear weapons are nuclear weapons themselves. But, if India's security requirements can be met through a universally acceptable and verifiable disarmament regime, India need not hesitate to disarm along with the others. In fact, to do so would be in its larger interests and it is for this reason that it makes sense to continue efforts for the realisation of an NWFW.

The chapter is broadly divided into three sections. The first one examines some of the predominant trends in the contemporary mood in India on the issue of nuclear disarmament. The second section then argues why it is still in India's interests, despite the change in mood and the country's status as a NWS to strive for disarmament. The final section identifies some of the steps that may be taken down this path.

India and Nuclear Disarmament: A Change in Mood?

As indicated briefly in the introduction to this chapter, Indian governments have long been vocal supporters and proactive participants in the debate on universal nuclear disarmament. Until the late 1980s, there was a genuine desire to achieve an NWFW. The Rajiv Gandhi Action Plan[5], in fact, in 1988, was presented to the Third Special Session on Disarmament with a lot of expectations. This was a well-considered and comprehensive document that identified specific actions that could be realistically carried out in different stages over 22 years, while simultaneously leading towards the establishment of a comprehensive global security system to usher in not just an NWFW but also a more broad-based non-violent world order. Had this been accepted in 1988 and implemented over the last two decades, the world would have been only one year away from bidding farewell to nuclear arms in 2010.

However, the idea proved to be ahead of its time. Not only were the NWS not interested in disarmament, wedded as they were to a MAD Cold War relationship, even within the Indian neighbourhood, Pakistan was well on its way towards nuclear weapons. In fact, on his return from New York, Rajiv Gandhi was met with intelligence reports on Pakistan's achievements in this direction, with the help of China. Through the 1990s, the nuclear security environment only deteriorated for India. Pakistan became more and more belligerent in its sub-conventional conflict in Kashmir from 1989 onwards, deriving confidence from the fact

5. Reproduced as Appendix 6 of this book.

that its nuclear deterrent provided it immunity against an Indian military response, and non-proliferation pressures on the country grew with the conclusion of the Comprehensive Test Ban Treaty (CTBT) in 1996. Finally, before the window of opportunity closed in 1999, the date by when India, along with 43 other nations that had nuclear programmes for electricity production had been asked to ratify the CTBT, India decided to conduct its nuclear tests on May11 and 13, 1998.

Given the cold response meted out to the Indian initiatives on disarmament, a general disregard for the idea of elimination of nuclear weapons, a simultaneous increase in the salience of nuclear weapons in the nuclear strategies of the NWS, and the increasing regional nuclear threat perceptions, it is not surprising that within India too there should have been the emergence of a more voluble opinion questioning the wisdom of India's pursuit of an NWFW in the past and into the future.[6] Especially after going nuclear, several opined that the country must now concentrate on building a credible nuclear deterrent rather than wasting its energies and time on articulating the merits of nuclear disarmament.

A change in mood has come about as a result of two factors: one, because of cynicism at the lack of seriousness on the part of the NWS. No country has changed its nuclear doctrine in any significant way to reflect a desire for an NWFW. Rather, each state still maintains that nuclear weapons are central to its national security. Two, the growing feeling about the unfeasibility of the exercise of eliminating nuclear weapons, of the inability to tackle issues such as the logistics and costs involved in the dismantling of tens of thousands of weapons, accounting of every gram of fissile material recovered, effective verification mechanisms, etc. This sentiment echoed a US government document of 1995 in which it was stated, "Since it is impossible to uninvent nuclear

6. See the writings of C. Raja Mohan and Bharat Karnad.

weapons or to prevent clandestine manufacture of some numbers of them, nuclear weapons seem destined to be the centerpiece of US strategic deterrence for the foreseeable future."[7] Indeed, given the complexities involved and the overlap of issues that would practically require an intense churning of international relations, it appears far easier and less risky to keep trudging along the known path of nuclear deterrence.

Consequently, there is a far greater emphasis on realism, or a takeover of idealism by *realpolitik* in the Indian polity. Idealism has slowly been replaced by greater realism in external relations and foreign policy-making. In this world of *realpolitik*, the ideal of nuclear disarmament has taken a beating. Of course, it cannot be overlooked that the voices of the realists contesting the official position had existed even as far back as the early 1960s when, after India's defeat at the hands of the Chinese, and Beijing's demonstration of a nuclear capability, the pro-bomb lobby including a considered military assessment had actively advocated that India should develop its own nuclear weapons. But the reach of this viewpoint had remained limited. In fact, till 1998, the pro-and anti-bomb lobbies had shared room for manoeuvrability as the official Indian position remained one of nuclear ambiguity. The sense of idealism and the principled approach was adequately balanced by the belief that national security considerations demanded the availability of all options. Post-1998, however, idealism took a beating as other more *real* issues of deteriorating regional security environs surfaced. Hence, the security dimension of nuclear policy has assumed greater sway with a consequent de-emphasis on disarmament.

7. "Essentials of Post-Cold War Deterrence", in Hans M. Kristensen, *Nuclear Futures: Proliferation of Weapons of Mass Destruction and US Nuclear Strategy*, BASIC Research Report 98.2, March 1998.

It is against this backdrop that it becomes far easier to understand the present less than enthusiastic response that some of the recent international initiatives on nuclear disarmament have generated within the country. India looks at the recent voices arguing for nuclear disarmament, influential though they are, with far greater caution and much lesser expectations. The two articles by the four US former secretaries in the *Wall Street Journal,* in January 2007 and January 2008, seem to have evoked some interest in the subject around the world. Soon thereafter, the Australian-Japan Bilateral Commission was set up to look at the issues involved. A Global Zero initiative has been launched in the US by a powerful Non-Governmental Organisation (NGO) and the UK is seriously considering the challenges related to verification.

India considers these as welcome moves, but its own experience in the matter generates a distinct sense of *déjà vu*. Emerging just 2-3 years ahead of the forthcoming NPT Review Conference (RevCon) in 2010, there is an opinion in India that contends that these may be targeted at showcasing some activity on nuclear disarmament in order to improve the credentials of the nuclear weapon states at the RevCon. In fact, these may even act as sweeteners with which to push more bitter pills of restricting the right of nations to the entire fuel cycle down the throats of non-nuclear weapon states. The real intent of the ongoing flurry of activity on nuclear disarmament will, therefore, be known by the middle of next year. The point to note, however, is that if the initiatives do fizzle out post- 2010, they would have done more harm to the cause of disarmament than if they had not emerged in the first place. This is because every time some such attempt is made, and it fails to meet with success, it further dents the chances of success of future such initiatives by reinforcing the mindset that an NWFW is not attainable.

And yet, India cannot afford to give up on its desire for an NWFW, however unfeasible it may seem at the moment. The following section explains why the realisation of such an end state contributes to India's security, and, hence, must be striven for as one strand of its nuclear strategy.

India's Stake in Nuclear Disarmament

India has long campaigned for the universal elimination of nuclear weapons on the grounds of morality and humanitarianism. While this logic shall hold eternally, in the present world of *realpolitik*, it would be worthwhile to argue the case for universal nuclear disarmament on the grounds of India's national security. In fact, since the decision to opt for nuclear weaponisation was dictated as a national security imperative, it is necessary that the decision to disarm must also be based on the same logic. Only if the country's national security is not jeopardised, but is rather enhanced, then and only then, can a persuasive case be made for India's continued pursuit of the objective of nuclear disarmament. And, this is precisely the point that this chapter makes in the paragraphs that follow. Of course, it need be said that India can accept only *universal* nuclear disarmament since it is only in such an environment that its national security would be enhanced. Unilateral nuclear disarmament is a non-option.

As has been argued earlier in the chapter, India's decision to overtly demonstrate its nuclear weapons capability was necessitated by a deteriorating regional security environment. The overt (China) and covert (Pakistan) presence of nuclear weapons in the immediate neighbourhood, and their nuclear and missile proliferation nexus made it imperative for India to acquire the same so as to hold forth an effective deterrent. But, if nuclear weapons can be removed from the neighbourhood, and at the global level, courtesy a universal nuclear disarmament regime, the situation would enhance the country's security. In fact, it is the presence of nuclear weapons that complicates India's security calculations.

As has been explained in the chapter on Pakistan's nuclear strategy, Islamabad makes good use of its nuclear weapons to deter not just India's nuclear weapons, but to checkmate the possibility of India using its conventional military superiority in response to its blatant sub-conventional conflict. Pakistan's quest and acquisition of a nuclear weapons capability was clearly driven by a desire to offset this superiority. With its nuclear weapons, Pakistan believes it has tied India's hands and curtailed its ability to use its conventionally superior force by projecting a risk that any war would quickly and automatically spiral into a nuclear confrontation. Premised on this logic, Islamabad is convinced that its perpetration of cross-border terrorism on Indian soil could continue unchecked since India's retaliatory options are limited. There is merit in this Pakistani assumption because India has certainly been compelled to review its retaliatory options and look for ways of fighting war without breaching Pakistan's perceived nuclear threshold. Of course, there is scope for a limited war even in the presence of nuclear weapons, as the last chapter has indicated; the conduct of this warfare, nevertheless, does impose limitations and yet can never be free of the fear of an inadvertent escalation to some sort of nuclear use.

The consequences of a nuclear exchange between India and Pakistan of only a few nuclear weapons of low yields of the kinds used in Hiroshima (about 15 kt) and Nagasaki (about 20 kt) could render fatalities and casualties in millions of lives and cause destruction of unimaginable dimensions and proportions. In fact, the death and destruction in the region, for several reasons (density of population being the most obvious, but only one of these), would be many orders of magnitude more than what the world has witnessed in the only nuclear use until now. It may not be forgotten that the nuclear damage wrought on Hiroshima and Nagasaki, for all its horrors, was *much less than what it could have been* for the simple reason that the weapons were not used at the optimal height in the case of Hiroshima, and the topography of Nagasaki helped the city escape far greater destruction. The Little Boy was dropped at a height, which resulted in a high airburst. Militaries are today aware that a weapon of similar yield, used as ground burst (at an optimal height of about 180 metres), would contaminate large sections of the target area and effectively render them uninhabitable for years. The fireball created by the ground burst would suck in a lot of dust, sand and other material and the radioactive fallout would then spread over a much larger area. Besides, of course, the weapon would have caused direct destruction in death and injuries through gamma and neutron radiation, blast overpressure and winds, and thermal fires over several kilometres around ground zero.

The above-mentioned facts should certainly influence the thinking of rational decision-makers. In fact, Pakistan, for all its nuclear bluster, would find it extremely difficult to actually use the weapon against India without creating a far more unfavourable and life-threatening situation for itself. And India, for all the tardiness and hesitation of action that it normally betrays, would find its rather high level of tolerance severely breached, resulting in punitive retaliatory action.

Besides the risk of a conventional conflict escalating into a nuclear exchange, the chances of which are relatively small, the presence of nuclear weapons also carries the threat of a nuclear war brought on by an unintended accident or miscalculation. In fact, modern day communication and sophisticated delivery systems take pride in extremely short lead times between the political decision to use nuclear weapons and the ability to bring it down on the pre-designated target. This carries the inherent risk of a nuclear weapon being launched during a crisis without adequate thought or in the heat of the moment. Neither can the danger of unauthorised nuclear launch be dismissed nor of the possibility of Pakistan's nuclear weapons slipping out of the control of the legitimate authority as a result of a long running state of political instability degrading the control of institutionalised structures. In fact, as long as nuclear weapons and terrorism cohabit in Pakistan, India cannot afford to take the threat of nuclear terrorism lightly. There are many routes of entry into India that terrorists could exploit to bring fissile material for an RDD or even a weapon—in a container from the sea; across land borders from any of India's neighbours; or even by a civilian aircraft from any Pak-friendly country. Difficult these acts may be, but they are not impossible. With state complicity, they could be carried out quite easily by non-state actors even as Pakistan may keep its facade of "plausible deniability".

It is for this reason that the elimination of nuclear weapons makes sense. If nuclear weapons were to be removed from national arsenals, India's conventional superiority would stand it a better chance to effectively respond to Pakistan's sub-conventional conflict. Rather, Pakistan would not be able to pursue its strategy of offensive defence. Nor would there be a risk of accidental, unauthorised or even deliberate nuclear use by radical elements in power or in collusion with the non-state actors.

As regards China, it may be said that India does not need nuclear weapons to meet any security challenge that it may pose. In fact, with a consistent approach of military modernisation, a non-nuclear China could not enjoy any substantial advantage in terms of its conventional strength vis-a-vis a non-nuclear India. Certainly, no nuclear coercion or blackmail would be possible by China. Meanwhile, in terms of actual war-fighting, in case the two countries were to confront one another with only conventional weaponry, there would be little for India to fear for several reasons. Firstly, given that the days of total war are virtually over, any war that the two might fight in the future would most likely be a limited one, in terms of scope, space and time i.e. it will be a limited, local, and essentially border war. In this context, if China were to fight a conventional war with India in order to redeem areas it claims in the northeast and further up in the disputed Himalayan reaches, then the terrain in which the hostilities would occur would provide an inherent advantage to India. Its lines of communication in the area are shorter and more secure. Even though China has invested a lot in building infrastructure on its side of the border to mitigate some of its vulnerabilities, India too can plug its own gaps through a well thought out and sustained effort in this direction. Maintaining a technological edge in Indian weapon systems, besides raising a few more mountain divisions, if required, could further strengthen this advantage. So, the point is that in the absence of a nuclear China, India could withstand and win a conventional onslaught, or at least deny victory to the adversary. Rather, it is the presence of nuclear weapons that tilts the balance in favour of China. It may be recalled that, as pointed out in Chapter 3 on China's nuclear strategy, an ongoing strategic modernisation would provide China the capability to graduate from minimum nuclear deterrence to limited nuclear deterrence, thereby enabling it to impose intra-war deterrence even after its first use of a nuclear weapon.

Therefore, a nuclear free India can more than match its nuclear free neighbours. An environment free of nuclear weapons would also obviate the risk of nuclear war brought on by an accidental, unauthorised or miscalculated use of nuclear weapons. This can be better explained through an analogy. When driving in chaotic conditions, one has to be responsible for not only one's own driving (which is relatively easy) but also remain constantly alert to the driving mistakes that others could make (which is infinitely more difficult) and which could endanger one's own security and existence. By the same logic, if nuclear weapons are available with others in the region, India must not only be able to develop and sustain an effective command and control system for its own weapons, but also hope that others too have foolproof measures in place. But, if there are no nuclear weapons with anyone, it would not only obviate the threat or the actuality of their deliberate use, but also eliminate any possibility of their inadvertent use through an accident or a miscalculation. This is even more crucial given that India, Pakistan and China lie in close vicinity to one another. Radioactive fallout from a nuclear explosion cannot be limited in space and the chance of some adverse effects on the nation that inflicts the misadventure cannot be completely ruled out. Given such facts, it seems logical that an NWFW would, as a happy by-product, ensure a safer regional security environment for India.

Identifying Milestones

This section of the chapter identifies some major milestones where the world could take a breather after trudging along on a journey that will be long and difficult. Long—because it is practically impossible to conceive that an NWFW could be achieved overnight. The very act of dismantlement and physical destruction of the large arsenals accumulated in the NWS would take time. And difficult—because the complexity of the problem is aggravated by the fact that it is intertwined with issues of state sovereignty, power play, national prestige, etc. A few caveats are, therefore, in order:

(a) An NWFW *cannot come about instantaneously*. The complexity of the task makes it impossible to expect that all countries would at some dawn see sense in arriving at a comprehensive treaty of nuclear abolition and do so in one go. It will have to be approached slowly, from different directions and with patience.

(b) An NWFW *can be achieved* if the agreed steps, however small, are taken with sincerity, backed by belief, sustained by passion, and supported by innovation. Any half-hearted measures that reveal lack of commitment would be harmful since they would reinforce the belief that it is an unfeasible proposition.

(c) An NWFW *need not wait* for all security-insecurity co-relations between countries to be addressed through a system of world governance in which all live in complete harmony before it can be realised. While a world that has renounced armed conflict as a means of settling disputes would indeed offer the ultimate security guarantee and remove justification for such weapons, the challenge, however, lies in conceiving nuclear abolition in this real world inhabited as it is by sovereign states with conflicting national interests.

(d) An NWFW must, therefore, follow a *mop-where-you-can* approach to begin with. Small, doable steps taken along a range of issues to address concerns and build confidence will enable more difficult steps to be taken in the years to come. Therefore, the approach must be calibrated to suit present conditions and taken in as cooperative a manner as possible.

One credible approach towards the goal of an NWFW exists in *progressively devaluing nuclear weapons and eventually delegitimising them.* It is human nature to cling on to anything that is perceived to be valuable. In case nuclear weapons can somehow be made worthless, nations would be more open to giving them up. Therefore, a devaluation strategy that deprives the weapons of utility and renders them unusable through a series of measures can prepare the ground for their eventual elimination.

Steps of devaluation may be taken along three trajectories:

attitudinal – relating to mindsets; actual – relating to doable acts; and accessorial – relating to additional measures. These need not be sequential. While each would create an environment making it easy to take another, any one of them could follow the other, or even be pursued simultaneously. The purpose would be to soften the mud around the deep-rooted tree of nuclear deterrence so that whenever the states are ready, it is possible to uproot it with ease.

Attitudinal Devaluation

Targeting Nuclear Belief Systems

Changing mindsets is imperative for devaluing nuclear weapons since the power of the mind cannot be underestimated. Since the majority of the strategic decision-makers have believed in the undesirability and impracticality of nuclear disarmament, this belief has gone on becoming entrenched. In order to break the mould, the belief systems need to be strategically "attacked" and replaced by those more favourable to nuclear abolition.

History provides ample examples of institutions whose abolition was once thought to be unthinkable but which broke down under the pressure of changed belief systems. One example exists in the case of abolition of slavery. In fact, when a few had started demanding its abolition, there were several who argued against it on the grounds that the economic and social systems would not be able to survive the drastic change and collapse!! At the time, on December 1, 1862, President Abraham Lincoln, in his famous message to the US Congress, had said, "The dogmas of the quiet past are inadequate to the stormy present… As our case is new, so we must think anew and act anew. We must disenthrall ourselves."[8]

8. Jerome B. Wiesner, "A Militarised Society," in Len Ackland and Steve McGuire, eds., *Assessing the Nuclear Age* (Chicago : Educational Foundation for Nuclear Science, 1986), p. 227.

This advice is most relevant today in the case of nuclear weapons. We have to disenthrall ourselves from their hold and if anything can do that it is the force of new ideas. There is an urgent need for brave, new thinkers who can visualise a new world and inject the passion for it into the decision-makers. Administrations are known to resist change but if a sufficiently enlightened political leadership can coalesce to form a critical mass, then the tipping point for nuclear disarmament, at least conceptually, can come about very quickly.

Redrafting Nuclear Doctrines to Restrict the Role of Nuclear Weapons

Altered belief systems would automatically prompt a re-examination and redrafting of nuclear doctrines. This would be an extremely important step since the role of a nuclear weapon is defined by a country's nuclear doctrine. To first restrict the role to a narrow core mission, and then take even that away, a rewriting of national nuclear doctrines is required. Nuclear weapons were originally developed to deter the use or threat of use of other nuclear weapons. Gradually, however, countries detected in them the power to offset even conventional inferiority. It is this belief that leads to the weapon being considered as one "of the weak against the strong, as the only weapon that can counter the conventional superiority of the West."[9] Not surprisingly, therefore, smaller states with heightened threat perceptions see value in nuclear weapons. Secondly, these weapons are also perceived as bestowers of prestige and status, as also the right to independent decision-making. In the words of the French Prime Minister Alain Juppe, "By acquiring nuclear capability, France was able to play, on the world scene, a role well above that justified by its mere

9. Robert G. Joseph, "Nuclear Deterrence and Regional Proliferators," *Washington Quarterly*, vol. 20, no. 3, Summer 1997, p. 168.

quantitative significance."[10] These perceptions enhance the utility of nuclear weapons and motivate others to strive for them. If more countries have to be stopped from perceiving a similar multi-role capability in nuclear weapons, then it is imperative to devalue the weapon by narrowing down its role.

Unfortunately though, today, all NWS have concluded that nuclear deterrence is still required as a principal pillar of their national security. Articulations such as these provide a positive feedback to nuclear deterrence and reinforce its hold over strategic thinking, making it a self-perpetuating phenomenon. It is this feedback loop that needs to be broken in order to devalue nuclear weapons. For this, a reassessment of the nuclear doctrines to make nuclear weapons less attractive by restricting, and eventually obviating their role in international politics becomes imperative.

Investing Confidence in Multilateral Treaties

The current mood on multilateral instruments is not very inspiring. The CTBT has been in a state of limbo from 1999 onwards. In 2002 once the US walked out of the ABM Treaty, Russia announced its abandonment of START II. Recent Russian pronouncements on the INF Treaty have put it under a cloud even as the FMCT is still awaiting direction in the Conference on Disarmament. All of these indicate a general trend away from arms control, particularly of the multilateral variety.

An expression of lack of faith in multilateral initiatives generates international instability, lowers confidence levels and increases the value of nuclear weapons. China, for example, has used every opportunity to decry the US attempt at seeking "absolute security" by freeing itself of international obligations, even as it undertakes

10. Speech delivered by French Prime Minister Alain Juppe at the *Institut des Hautes Etudes de Defence Nationale*, Paris on September 7, 1995, and as reproduced in *Strategic Digest*, April 1998.

rapid military modernisation that would include an increase in its own nuclear arsenal, development of counter-measure and space-based capabilities, and thereby raises the concerns of others. In this process, then, international security and stability are the losers as the value of nuclear weapons soars.

Actual Devaluation

Accepting No First Use (NFU)

While limiting the role of nuclear weapons to dealing with only nuclear contingencies provides negative security assurances (NSA) to NNWS and devalues the weapon to some extent, accepting NFU takes the task further by delegitimising their use against NWS too. Therefore, adoption of NFU would be a crucial step towards the eventual delegitimisation of nuclear weapons since it would involve an assurance from every country that it would not be the first to introduce nuclear weapons into a conflict. Since there will not be a first, it would effectively mean no use of the nuclear weapon. If the NFU could be implemented through a legally binding, universally accepted agreement, then any first use would constitute a violation of international law, inviting crippling universal sanctions, worldwide opprobrium and penalties that could be consensually agreed to.

Of course, there are critics of no first use who dismiss it as nothing more than a declaratory policy that would mean little once hostilities actually break out between two nuclear nations. Such criticism, however, tends to overlook two facts. Firstly, that the adoption of NFU automatically translates into a different nuclear force posture and deployment pattern that ensure that the promise of no first use is kept. Doctrines that ascribe a war-fighting role to nuclear weapons envisage 'first use' to retain the military advantage and adopt launch-on-warning (LOW) or launch-under-attack (LUA) postures. To undertake preemption, both sides need a large infrastructure in the form of C3I, early warning, defences, etc. NFU, on the other hand, frees the nation of these requirements by basing its nuclear strategy on a retaliation only policy. It allows for greater response time and a more relaxed posture, thereby also relieving pressure on the adversary.

Secondly, the efficacy of NFU flows from the fact that declarations on the conduct of war do have relevance and cannot be summarily dismissed. These laws set certain limits to the levels of permissible violence and establish useful norms. For instance, the Geneva Protocol of 1925 outlawed the use of chemical and biological weapons, and except for rare exceptions, these weapons have remained out of the realm of war-fighting. In fact, eventually, though several decades down the line, it did become possible to capitalise upon the norm to formalise conventions outlawing these WMD. Following this example, an NFU commitment from all NWS could be expected to translate into a Nuclear Weapons Convention some time in the future.[11]

A universal NFU would be even more relevant as nuclear weapons reduce. With small nuclear forces, the temptation to launch a disarming first strike would be high because of the 'use them or lose them' compulsions. But an NFU posture would remove this temptation for self and the adversary. If the adversary is under constant fear that a nuclear strike is imminent, its own temptation to use nuclear force would be higher. But if all NWS were to accept the NFU, then none

11. It may be recalled that the Geneva Protocol of June 17, 1925, also started out rather tentatively. It placed a prohibition on the use in war of chemicals and bacteriological methods of warfare, but did not stop nations from developing, producing or stockpiling these weapons. Not all nations became party to the Protocol and some that did, like the UK, reserved the right to use such weapons in retaliation.

would ever initiate a nuclear strike. Lengthening the fuse on their possible use would be a step towards downgrading the importance of nuclear weapons since military planners would have to discount their role in contingency planning. Over a period of time then, the weapons could fall into a pattern of disuse and lose their utility.

Banning the Use or Threat of Use of Nuclear Weapons

A step even beyond the NFU, an agreement prohibiting the very use or threat of use of nuclear weapons would provide another meaningful way to delegitimise the weapon. A taboo against nuclear use already exists. If this can be legalised, it would further devalue the weapon. In case all NWS were to make a solemn declaration that nuclear weapons shall not be used and that any country using them or threatening to use them, shall face commensurate retribution and a total boycott, it would significantly devalue nuclear weapons. None would want to acquire a weapon that could not be used, not in war, and, hence, not as a deterrent either. Therefore, a total ban on the use of nuclear weapons would directly strike at the very root of their utility.

In this context, it might be mentioned that a norm of nuclear non-use already exists. Widespread popular revulsion against nuclear weapons and widely held inhibitions on their use influence nuclear decision-making. This was evident even in the 1950s when an officer in the State Department's Bureau of Far East Affairs warned, during the Korean War, that "even though the military results achieved by atomic bombardment may be identical to those attained by conventional weapons, the effect on world opinion will be vastly different. The A-bomb has the status of a peculiar monster conceived by American cunning and its use by us in whatever situation would be exploited to our serious detriment."[12] Since then, the norm has existed but it has never been legalised.

In fact, the legality of nuclear use remains in dispute because the 1996 International Court of Justice Advisory opinion had announced that while the use or threat of use is "generally" unlawful, it could not "definitively" conclude whether it would be unlawful in extreme circumstances of self-defence if the survival of the state was at stake. The ambiguity is, therefore, maintained. Meanwhile, the UN General Assembly has periodically considered resolutions stating that the use of nuclear weapons was contrary to the "spirit, letter and aims of the UN." Way back in 1961, it had adopted a declaration by a vote of 55 to 20 with 26 abstentions. The US and NATO had then opposed it, contending that in the event of aggression, the attacked nation should be free to take whatever action with whatever weapons not specifically banned by international law. India, for several years now, has proposed a "Convention on the Prohibition of Use of Nuclear Weapons". It seeks a multilateral, universal and binding agreement prohibiting the use or threat of use of nuclear weapons through an international convention. In the year 2000, however, the move was able to muster only 109 votes in favour, with 43 against and 16 abstentions. Predictably, the P-5 opposed the resolution. The US hailed instead a step-by-step process that embraced unilateral, bilateral and multilateral measures. [13] Interestingly, Japan, for all its abhorrence of nuclear weapons, also found it prudent to abstain for reasons similar to those voiced by the US. Such country positions, nevertheless, underestimate the significance of actually arriving

12. Nina Tannenwald, "The Nuclear Taboo: The United States and the Normative Basis of Nuclear Non-use", *International Organisation,* vol.53. no.3, Summer 1999, pp. 433-468.

13. See "Appendix: Summary of Resolutions",*Disarmament Diplomacy* <http://www.acronym.org>

at a convention banning the use of nuclear weapons to devalue them substantially.

This is even more urgent today because the norm of non-use or the taboo against the use of nuclear weapons could erode in the future for, at least, three reasons. One, because the memory of the horrors of Hiroshima and Nagasaki is fading. The new generation that has no first-hand experience of the damage and destruction that a nuclear weapon can cause may not be able to visualise the implications of nuclear use. Second, with the induction of mini or micro nuclear weapons, the temptation to use them on the assumption that they would cause minimal collateral damage cannot be discounted. The convergence of the destructive power of small nuclear weapons and advanced conventional weapons, as in the case of fuel air explosives would also erode the distinction between the two. This could lead to a casual resort to nuclear weapons if they come to be perceived as no different than other weapons. Third, there could be forces that have a vested interest in destroying the taboo. For instance, it is documented that even in 1953, in a discussion with the US Presiden, Secretary of State John Foster Dulles had admitted that "since in the present state of world opinion we could not use an A-bomb, we should make every effort now to dissipate this feeling, especially since we are spending such vast sums on the production of weapons we cannot use." There is nothing to suggest that similar opinions are not being, or will not be, voiced in the future. In view of these, a ban on nuclear use would immediately serve to remove the value of nuclear weapons. It is no longer possible to put the nuclear genie back, but it is still possible that the genie of nuclear use is not allowed to come out of the bottle.

Halting Further R&D for Nuclear Modernisation

Besides reducing the existing nuclear arsenals, it is also imperative that a complete halt be put to further modernisation of nuclear weapons. This is particularly important given the emergence of a new trend towards improving the weapons in such a way as to make them "more user-friendly." In the US, certain laboratories have pushed for the development of mini-nukes of smaller yields. It is contended that small, low-yield nuclear weapons could be effectively used to destroy even a deeply buried or hardened underground facility without causing too much collateral damage.[14] In his testimony before the Senate Armed Services Sub-Committee Linton Brooks, Administrator of the NNSA, said, "Today's Cold War legacy stockpile is the wrong stockpile from a number of perspectives"[15] because it contained too many high yield weapons that offered little or no capability against hard and deeply buried targets, provided no way to limit collateral damage, and could not destroy chemical or biological munitions.

Any decision validating the importance of smaller, low-yield and more usable nuclear weapons sets off a dangerous chain reaction with global ramifications. Russia, is already believed to be thinking along similar lines. In mid-December 2006, Putin called for the "development of cutting edge strategic weapons" and emphasised "quality over quantity".[16] Such developments ascribe value to nuclear weapons instead of taking it away from them. Until now arms control has focussed on the quantitative growth of arsenals. Disarmament must devise arrangements for controlling the continuous qualitative upgradation of nuclear and conventional weapons.

---

14. Theresa Hitchens, "The US Nuclear Debate: Issues of Concern," *BASIC Paper* no. 35, February 2001.
15. Statement at cited in Stephen I. Schwartz, "Warheads Aren't Forever," *Bulletin of Atomic Scientists*, vol 61, no. 5, September/October 2005, pp. 58-64.
16 "Russia Against US MD Plans for Europe", http://www.spacewar.com, December 27, 2006.

Accessorial Devaluation

Crafting a Safeguards Regime

In order for the measures outlined above to be meaningful and sustainable, it becomes necessary to craft an effective and acceptable safeguards regime that can provide adequate guarantee that nations are complying with their commitments. Such a regime would be possible only through the adept employment of a combination of technological, diplomatic and statecraft skills. It would have to trudge a fine line because in order to be effective, it would have to be relatively intrusive and in order to be acceptable it would have to be relatively limited. The organisational and institutional mechanisms that would perform these functions would also need to be thought through. Of course, the IAEA that has several decades of experience in safeguards is an option. But it would need to be infused with a fresh flow of funds and political powers to take on the additional burden of new safeguards, which would arise after a careful definition of items, activities, and facilities that would need to be monitored.

Joint R & D of Verification Technologies

States can be expected to renounce the nuclear weapon only when enough technical expertise is available to ensure adequate verification of the commitments undertaken to abandon nuclear arsenals. History illustrates that often changes become acceptable only when it becomes technically possible to prove something. For centuries, man continued to believe that the sun revolves around the earth. Those who disputed this belief were ridiculed or persecuted. Copernicus in his lifetime could not find acceptance for his theory. The belief changed only after the invention of the telescope and other astronomical instruments that could empirically prove that the sun was actually the centre of the universe. Applying the same logic to nuclear disarmament, it may be said that advancements in the field of surveillance and monitoring technologies coupled with the research and development on how to deal with the nuclear materials obtained from dismantled weapons would help overcome the technical challenges to feasibility of disarmament. If pursued as a multinational effort, it could help build deeper technical relations and lead to greater confidence in the process.

Dealing with Delinquents

In order to devalue nuclear weapons, it becomes necessary to put in place effective enforcement strategies that clearly lay down penalties for cheating or breakout. The guarantee of a collective and concerted international response could act as an effective deterrent to the clandestine development of nuclear weapons once all countries have committed to their non-use or abolition. A resolute response with sure penalties for any breach of trust would stem the temptation to go down the nuclear path or to use it as a lucrative bargaining chip. Only a collective endeavour to respond surely and unitedly could effectively serve to devalue nuclear weapons. Of course, a determination of the exact nature of penalties that may be acceptable to all requires further investigation. But it is not a challenge that cannot be surmounted if the requisite political will exists.

Conclusion

The chapter has presented its case for why the pursuit of universal nuclear disarmament makes for a good nuclear strategy for India. Even though there is far less enthusiasm, or perhaps a greater sense of realism on its prospects, it must, nevertheless remain an aspiration since it could resolve several security challenges for the country. It is with this in view that the chapter has also identified a number of steps that could help to devalue the importance of nuclear weapons. These are largely based on common sense and logic. However, country positions are not always dictated by these two parameters. Ample evidence of this is provided by national positions as revealed in the voting patterns on the different resolutions that continue to be placed before the United Nations General Assembly (UNGA) on the issue of disarmament. In the year 2000 (as also every year on an average) nearly four dozen such resolutions appeared before the UN. However, an examination of the voting patterns revealed the rather predictable positions of each country. The NWS and their allies tend to support resolutions listing the progress already made in the direction in terms of the increasing membership of the NPT, the conclusion of the CTBT, negotiations on the FMCT, reductions negotiated under the START process etc. However, they oppose all resolutions seeking an actual delineation of steps to eliminate nuclear weapons in a time-bound or a phased manner.

However, nuclear weapons can be abolished only when the belief systems behind their use and utility change. Thereafter, these changes can be institutionalised as norms and in legal regimes. Initially, it shall all have to begin in the mind and thereafter the possibilities would grow. Of course, one cannot set a date to when these latent possibilities are realised. Every revolution in history has taken its leaders by surprise because as long as the old order exists, it tends to hide its fragility through elaborate measures. Its control then seems total and impregnable. But ideas can still sneak in and slowly snowball into a force capable of bringing about a profound change.

Similarly, the vision of an NWFW too can be realised if it can draw upon new thinking at different levels: exemplary political vision and statesmanship willing to discard entrenched beliefs and accept new norms of inter-state behaviour; resolute military courage to shake off the crutch of nuclear deterrence and accept seemingly lower levels of security, especially for states that are conventionally weaker and see nuclear weapons as levellers of sorts; large doses of logistic innovation to deal with the discarded weapons; an international system based on greater transparency, confidence and more participative institutionalised mechanisms.

Complete elimination has often been dismissed as unattainable unless war as a means of settling disputes between states is first abolished and a state of general and complete disarmament is enforced through a system of world governance. Since the attainment of these prerequisites appears Utopian, nuclear disarmament too tends to be dismissed as unattainable. However, simple and easily implementable steps to devalue nuclear weapons prove that there are worthwhile ways of approaching nuclear abolition.

Appendix 1

Draft Report of National Security Advisory Board on Indian Nuclear Doctrine

August 17, 1999

1. Preamble

1.1. The use of nuclear weapons in particular as well as other weapons of mass destruction constitutes the gravest threat to humanity and to peace and stability in the international system. Unlike the other two categories of weapons of mass destruction, biological and chemical weapons which have been outlawed by international treaties, nuclear weapons remain instruments for national and collective security, the possession of which on a selective basis has been sought to be legitimized through permanent extension of the Nuclear Non-Proliferation Treaty (NPT) in May 1995. Nuclear weapon states have asserted that they will continue to rely on nuclear weapons with some of them adopting policies to use them even in a non-nuclear context. These developments amount to virtual abandonment of nuclear disarmament. This is a serious setback to the struggle of the international community to abolish weapons of mass destruction.

1.2. India's primary objective is to achieve economic, political, social, scientific and technological development within a peaceful and democratic framework. This requires an environment of durable peace and insurance against potential risks to peace and stability. It will be India's endeavour to proceed towards this overall objective in cooperation with the global democratic trends and to play a constructive role in advancing the international system toward a just, peaceful and equitable order.

1.3. Autonomy of decision-making in the developmental process and in strategic matters is an inalienable democratic right of the Indian people. India will strenuously guard this right in a world where nuclear weapons for a select few are sought to be legitimized for an indefinite future, and where there is growing complexity and frequency in the use of force for political purposes.

1.4. India's security is an integral component of its development process. India continuously aims at promoting an ever-expanding area of peace and stability around it so that developmental priorities can be pursued without disruption.

1.5. However, the very existence of offensive doctrines pertaining to the first use of nuclear weapons and the insistence of some nuclear weapons states on the legitimacy of their use even against non-nuclear weapon countries constitute a threat to peace and stability.

1.6. This document outlines the broad principles for the development, deployment and employment of India's nuclear forces. Details of policy and strategy concerning force structures, deployment and employment of nuclear forces will flow from this framework and will be laid down separately and kept under constant review.

2. Objectives

2.1. In the absence of global nuclear disarmament, India's strategic interests require effective, credible nuclear deterrence and adequate retaliatory capability should deterrence fail. This is consistent with the UN Charter, which sanctions the right of self-defence.

2.2. The requirements of deterrence should be carefully weighed in the design of Indian nuclear forces and in the strategy to provide for a level of capability consistent with maximum credibility, survivability, effectiveness, safety and security.

2.3. India shall pursue a doctrine of credible minimum nuclear deterrence. In this policy of "retaliation only", the survivability of our arsenal is critical. This is a dynamic concept related to the strategic environment, technological imperatives and the needs of national security. The actual size components, deployment and employment of nuclear forces will be decided in the light of these factors. India's peacetime posture aims at convincing any potential aggressor that :

(a) any threat of use of nuclear weapons against India shall invoke measures to counter the threat: and

(b) any nuclear attack on India and its forces shall result in punitive retaliation with nuclear weapons to inflict damage unacceptable to the aggressor.

2.4. The fundamental purpose of Indian nuclear weapons is to deter the use and threat of use of nuclear weapons by any State or entity against India and its forces. India will not be the first to initiate a nuclear strike, but will respond with punitive retaliation should deterrence fail.

2.5. India will not resort to the use or threat of use of nuclear weapons against States which do not possess nuclear weapons, or are not aligned with nuclear weapon powers.

2.6. Deterrence requires that India maintain:

(a) Sufficient, survivable and operationally prepared nuclear forces,

(b) a robust command and control system,

(c) effective intelligence and early warning capabilities,

(d) comprehensive planning and training for operations in line with the strategy, and

(e) the will to employ nuclear forces and weapons

2.7. Highly effective conventional military capabilities shall be maintained to raise the threshold of outbreak both of conventional military conflict as well as that of threat or use of nuclear weapons.

3. Nuclear Forces

3.1. India's nuclear forces will be effective, enduring, diverse, flexible, and responsive to the requirements in accordance with the concept of credible minimum deterrence. These forces will be based on a triad of aircraft, mobile land-based missiles and sea-based assets in keeping with the objectives outlined above.

Survivability of the forces will be enhanced by a combination of multiple redundant systems, mobility, dispersion and deception.

3.2. The doctrine envisages assured capability to shift from peacetime deployment to fully employable forces in the shortest possible time, and the ability to retaliate effectively even in a case of significant degradation by hostile strikes.

4. Credibility And Survivability

The following principles are central to India's nuclear deterrent

4.1. **Credibility:** Any adversary must know that India can and will retaliate with sufficient nuclear weapons to inflict destruction and punishment that the aggressor will find unacceptable if nuclear weapons are used against India and its forces.

4.2. **Effectiveness***:* The efficacy of India's nuclear deterrent to be maximized through synergy among all elements involving reliability, timeliness, accuracy and weight of the attack.

4.3 **Survivability***:*

(i) India's nuclear forces and their command and control shall be organized for very high survivability against surprise attacks and for rapid punitive response. They shall be designed and deployed to ensure survival against a first strike and to endure repetitive attrition attempts with adequate retaliatory capabilities for a punishing strike which would be unacceptable to the aggressor.

(ii) Procedures for the continuity of nuclear command and control shall ensure a continuing capability to effectively employ nuclear weapons.

5. Command And Control

5.1. Nuclear weapons shall be tightly controlled and released for use at the highest political level. The authority to release nuclear weapons for use resides in the person of the Prime Minister of India, or the designated successor(s).

5.2. An effective and survivable command and control system with requisite flexibility and responsiveness shall be in place. An integrated operational plan, or a series of sequential plans, predicated on strategic objectives and a targeting policy shall form part of the system.

5.3. For effective employment, the unity of command and control of nuclear forces, including dual capable delivery systems shall be ensured.

5.4. The survivability of the nuclear arsenal and effective command, control, communications, computing, intelligence and information (C4I2) systems shall be assured.

5.5. The Indian defence forces shall be in a position to, execute operations in an NBC environment with minimal degradation;

5.6. Space-based and other assets shall be created to provide early warning, communications, damage/detonation assessment.

6. Security and Safety

6.1. **Security:** Extraordinary precautions shall be taken to ensure that nuclear weapons, their manufacture, transportation and storage are fully guarded against possible theft, loss, sabotage, damage or unauthorized access or use.

6.2. **Safety** is an absolute requirement and tamper proof procedures and systems shall be instituted to ensure that unauthorized or inadvertent activation/use of nuclear weapons does not take place and risks of accident are avoided.

6.3. **Disaster control:** India shall develop an appropriate disaster control system capable of handling the unique requirements of potential incidents involving nuclear weapons and materials;

7. Research and Development

7.1. India should step up efforts in research and development to keep up with technological advances in this field.

7.2. While India is committed to maintain the deployment of a deterrent which is both minimum and credible, it will not accept any restraints on building its R&D capability.

8. Disarmament And Arms Control

8.1. Global, verifiable and non-discriminatory nuclear disarmament is a national security objective. India shall continue its efforts to achieve the goal of a nuclear weapon-free world at an early date.

8.2. Since no first use of nuclear weapons is India's basic commitment, every effort shall be made to persuade other States possessing nuclear weapons to join an international treaty banning first use.

8.3. Having provided unqualified negative security assurances, India shall work for internationally binding unconditional negative security assurances by nuclear weapon states to non-nuclear weapon states.

8.4. Nuclear arms control measures shall be sought as part of national security policy to reduce potential threats and to protect our own capability and its effectiveness.

8.5. In view of the very high destructive potential of nuclear weapons, appropriate nuclear risk reduction and confidence building measures shall be sought, negotiated and instituted.

Source: http://meaindia.nic.in/

Appendix 2

The Cabinet Committee on Security Reviews Operationalization of India's Nuclear Doctrine

New Delhi
January 4, 2003

The Cabinet Committee on Security (CCS) met today to review the progress in operationalizing of India's nuclear doctrine. The Committee decided that the following information, regarding the nuclear doctrine and operational arrangements governing India's nuclear assets, should be shared with the public.

2. India's nuclear doctrine can be summarized as follows:

(i) Building and maintaining a credible minimum deterrent;

(ii) A posture of "No First Use": nuclear weapons will only be used in retaliation against a nuclear attack on Indian territory or on Indian forces anywhere;

(iii) Nuclear retaliation to a first strike will be massive and designed to inflict unacceptable damage.

(iv) Nuclear retaliatory attacks can only be authorized by the civilian political leadership through the Nuclear Command Authority.

(v) Non-use of nuclear weapons against non-nuclear weapon states;

(vi) However, in the event of a major attack against India, or Indian forces anywhere, by biological or chemical weapons, India will retain the option of retaliating with nuclear weapons;

(vii) A continuance of strict controls on export of nuclear and missile related materials and technologies, participation in the Fissile Material Cut-off Treaty negotiations, and continued observance of the moratorium on nuclear tests.

(viii) Continued commitment to the goal of a nuclear weapon-free world, through global, verifiable and non-discriminatory nuclear disarmament.

3. The Nuclear Command Authority comprises a Political Council and an Executive Council. The Political Council is chaired by the Prime Minister. It is the sole body which can authorize the use of nuclear weapons.

4. The Executive Council is chaired by the National Security Advisor. It provides inputs for decision-making by the Nuclear Command Authority and executes the directives given to it by the Political Council.

5. The CCS reviewed the existing command and control structures, the state of readiness, the targeting strategy for a retaliatory attack, and operating procedures for various stages of alert and launch. The Committee expressed satisfaction with the overall preparedness. The CCS approved the appointment of a Commander-in-Chief, Strategic Forces Command, to manage and administer all Strategic Forces.
6. The CCS also reviewed and approved the arrangements for alternate chains of command for retaliatory nuclear strikes in all eventualities.

Source: http://meaindia.nic.in/

Indo-US Joint Statement Between the Prime Minister Dr. Manmohan Singh and the US President Mr. George W. Bush

July 18, 2005, Washington DC

The following is the text of the Indo-US Joint Statement issued after the delegation-level meeting, in Washington DC.

"Prime Minister Manmohan Singh and President Bush today declare their resolve to transform the relationship between their countries and establish a global partnership. As leaders of nations committed to the values of human freedom, democracy and rule of law, the new relationship between India and the United States will promote stability, democracy, prosperity and peace throughout the world. It will enhance our ability to work together to provide global leadership in areas of mutual concern and interest.

Building on their common values and interests, the two leaders resolve:

- To create an international environment conducive to promotion of democratic values, and to strengthen democratic practices in societies which wish to become more open and pluralistic.
- To combat terrorism relentlessly. They applaud the active and vigorous counterterrorism cooperation between the two countries and support more international efforts in this direction. Terrorism is a

global scourge and the one we will fight everywhere. The two leaders strongly affirm their commitment to the conclusion by September of a UN comprehensive convention against international terrorism.

The Prime Minister's visit coincides with the completion of the Next Steps in Strategic Partnership (NSSP) initiative, launched in January 2004. The two leaders agree that this provides the basis for expanding bilateral activities and commerce in space, civil nuclear energy and dual-use technology.

Drawing on their mutual vision for the US-India relationship, and our joint objectives as strong long-standing democracies, the two leaders agree on the following:

For the Economy

- Revitalize the US-India Economic Dialogue and launch a CEO Forum to harness private sector energy and ideas to deepen the bilateral economic relationship.
- Support and accelerate economic growth in both countries through greater trade, investment, and technology collaboration.
- Promote modernization of India's infrastructure as a prerequisite for the continued growth of the Indian economy. As India enhances its investment climate, opportunities for investment will increase.
- Launch a US-India Knowledge Initiative on Agriculture focussed on promoting teaching, research, service and commercial linkages.

For Energy and the Environment

- Strengthen energy security and promote the development of stable and efficient energy markets in India with a view to ensuring adequate, affordable energy supplies and conscious of the need for sustainable development. These issues will be addressed through the US-India Energy Dialogue.
- Agree on the need to promote the imperatives of development and safeguarding the environment, commit to developing and deploying cleaner, more efficient, affordable, and diversified energy technologies.

For Democracy and Development

- Develop and support, through the new US-India Global Democracy

Initiative in countries that seek such assistance, institutions and resources that strengthen the foundations that make democracies credible and effective. India and the US will work together to strengthen democratic practices and capacities and contribute to the new UN Democracy Fund.

- Commit to strengthen cooperation and combat HIV/AIDS at a global level through an initiative that mobilizes private sector and government resources, knowledge, and expertise.

For Non-Proliferation and Security

- Express satisfaction at the New Framework for the US-India Defense Relationship as a basis for future cooperation, including in the field of defense technology.
- Commit to play a leading role in international efforts to prevent the proliferation of Weapons of Mass Destruction (WMD). The US welcomed the adoption by India of legislation on WMD (Prevention of Unlawful Activities Bill).
- Launch a new US-India Disaster Relief Initiative that builds on the experience of the Tsunami Core Group, to strengthen cooperation to prepare for and conduct disaster relief operations.

For High-Technology and Space

- Sign a Science and Technology Framework Agreement, building on the US-India High-Technology Cooperation Group (HTCG), to provide for joint research and training, and the establishment of public-private partnerships.
- Build closer ties in space exploration, satellite navigation and launch, and in the commercial space arena through mechanisms such as the US-India Working Group on Civil Space Cooperation.
- Building on the strengthened non-proliferation commitments undertaken in the NSSP, to remove certain Indian organizations from the Department of Commerce's Entity List.

Recognizing the significance of civilian nuclear energy for meeting growing global energy demands in a cleaner and more efficient manner, the two leaders discussed India's plans to develop its civilian nuclear energy program.

President Bush conveyed his appreciation to the Prime Minister over India's strong commitment to preventing WMD proliferation and stated that as a responsible state with advanced nuclear technology, India should acquire the same benefits and advantages as other such states. The President told the Prime Minister that he will work to achieve full civil nuclear energy cooperation with India as it realizes its goals of promoting nuclear power and achieving energy security. The President would also seek agreement from Congress to adjust US laws and policies, and the United States will work with friends and allies to adjust international regimes to enable full civil nuclear energy cooperation and trade with India, including but not limited to expeditious consideration of fuel supplies for safeguarded nuclear reactors at Tarapur. In the meantime, the United States will encourage its partners to also consider this request expeditiously. India has expressed its interest in ITER and a willingness to contribute. The United States will consult with its partners considering India's participation. The United States will consult with the other participants in the Generation IV International Forum with a view toward India's inclusion.

The Prime Minister conveyed that for his part, India would reciprocally agree that it would be ready to assume the same responsibilities and practices and acquire the same benefits and advantages as other leading countries with advanced nuclear technology, such as the United States. These responsibilities and practices consist of identifying and separating civilian and military nuclear facilities and programs in a phased manner and filing a declaration regarding its civilians facilities with the International Atomic Energy Agency (IAEA); taking a decision to place voluntarily its civilian nuclear facilities under IAEA safeguards; signing and adhering to an Additional Protocol with respect to civilian nuclear facilities; continuing India's unilateral moratorium on nuclear testing; working with the United States for the conclusion of a multilateral Fissile Material Cut-off Treaty; refraining from transfer of enrichment and reprocessing technologies to states that do not have them and supporting international efforts to limit their spread; and ensuring that the necessary steps have been taken to secure nuclear materials and technology through comprehensive export control legislation and through harmonization and adherence to Missile Technology Control Regime (MTCR) and Nuclear Suppliers Group (NSG) guidelines.

The President welcomed the Prime Minister's assurance. The two leaders agreed to establish a working group to undertake on a phased basis in the months ahead the necessary actions mentioned above to fulfill these commitments. The President and Prime Minister also agreed that they would review this progress when the President visits India in 2006.

The two leaders also reiterated their commitment that their countries would play a leading role in international efforts to prevent the proliferation of weapons of mass destruction, including nuclear, chemical, biological and radiological weapons.

In light of this closer relationship, and the recognition of India's growing role in enhancing regional and global security, the Prime Minister and the President agree that international institutions must fully reflect changes in the global scenario that have taken place since 1945. The President reiterated his view that international institutions are going to have to adapt to reflect India's central and growing role. The two leaders state their expectations that India and the United States will strengthen their cooperation in global forums.

Prime Minister Manmohan Singh thanks President Bush for the warmth of his reception and the generosity of his hospitality. He extends an invitation to President Bush to visit India at his convenience and the President accepts that invitation."

Appendix 4

Implementation of the India-United States Joint Statement of July 18, 2005: India's "Separation Plan" tabled in Parliament on March 7, 2006

"The resumption of full civilian nuclear energy cooperation between India and the United States arose in the context of India's requirement for adequate and affordable energy supplies to sustain its accelerating economic growth rate and as recognition of its growing technological prowess. It was preceded by discussions between the two Governments, particularly between President Bush and Prime Minister Manmohan Singh, of the global energy scenario and the long-term implications of increasing pressure on hydrocarbon resources and rising oil prices. These developments led to the announcement in April 2005 of an Indo-US Energy Dialogue that encompassed the entire spectrum of energy options ranging from oil and gas to coal, alternative fuels and civilian nuclear energy. Through the initiation of a sustained dialogue to address energy security concerns, the two countries sought to promote stable, efficient, predictable and cost-effective solutions for India's growing requirements. At the same time, they also agreed on the need to develop and deploy cleaner, more efficient, affordable and diversified energy technologies to deal with the environmental implications of energy consumption. India had developed proven and wide-ranging capabilities in the nuclear sector, including over the entire nuclear fuel cycle. It is internationally recognized that India has unique contributions to make to international efforts towards meeting these objectives. India has become a full partner in ITER, with the full support of the US and other partners. India also accepted the US invitation to join the initiative on Clean Development Partnership.

US Undertaking

"2. Noting the centrality of civilian nuclear energy to the twin challenges of energy security and safeguarding the environment, the two Governments agreed on 18 July 2005 to undertake reciprocal commitments and responsibilities that would create a framework for the resumption of full cooperation in this field. On its part, the United States undertook to:

- Seek agreement from the Congress to adjust US laws and policies to achieve full civil nuclear energy cooperation.
- Work with friends and allies to adjust international regimes to enable full civil nuclear energy cooperation and trade with India, including but not limited to expeditious consideration of fuel supplies for safeguarded nuclear reactors at Tarapur.
- In the meantime, encourage its partners to consider fuel supply to Tarapur expeditiously.
- To consult with its partners to consider India's participation in ITER.
- To consult with other participants in the Generation-IV International Forum with a view towards India's inclusion.

"3. India had conveyed its readiness to assume the same responsibilities and practices and acquire the same benefits and advantages as other leading countries with advanced nuclear technology, such as the United States. Accordingly, India for its part undertook the following commitments:

- Identifying and separating civilian and military nuclear facilities and programmes in a phased manner.
- Filing a declaration regarding its civilian facilities with the IAEA.
- Taking a decision to place voluntarily its civilian nuclear facilities under IAEA safeguards, and
- Signing and adhering to an Additional Protocol with respect to civilian nuclear facilities.

"4. Other commitments undertaken by India have already been fulfilled in the last year. Among them are:

- India's responsible non-proliferation record, recognized by the US, continues and is reflected in its policies and actions.
- The harmonization of India's export controls with NSG (Nuclear Suppliers' Group) and MTCR (Missile Technology Control Regime) Guidelines even though India is not a member of either group. These guidelines and control lists have been notified and are being implemented.
- A significant upgrading of India's non-proliferation regulations and export controls has taken place as a result of the Weapons of Mass Destruction Act of May 2005. Inter-Ministerial consultations are ongoing to examine and amend other relevant Acts as well as framing appropriate rules and regulations.
- Refrain from transfer of enrichment and reprocessing technologies to states that do not have them and supporting international efforts to limit their spread. This has guided our policy on non-proliferation.
- Continued unilateral moratorium on nuclear testing, and
- Willingness to work with the United States for the conclusion of a multilateral Fissile Material Cut-off Treaty.

"5. The Joint Statement of July 18, 2005, recognized that India is ready to assume the same responsibilities and practices as other leading countries with advanced nuclear technology, such as the United States. India has an impeccable record in non-proliferation. The Joint Statement acknowledges that India's nuclear programme has both a military and a civilian component. Both sides had agreed that the purpose was not to constrain India's strategic programme but to enable resumption of full civil nuclear energy cooperation in order to enhance global energy and environmental security. Such cooperation was predicated on the assumption that any international civil nuclear energy cooperation (including by the US) offered to India in the civilian sector should, firstly, not be diverted away from civilian purposes, and secondly, should not be transferred from India to third countries without safeguards. These concepts will be reflected in the Safeguards Agreement to be negotiated by India with IAEA.

"6. India's nuclear programme is unique as it is the only state with nuclear weapons not to have begun with a dedicated military programme. It must be appreciated that the strategic programme is an offshoot of research on nuclear power programme and consequently, it is embedded in a larger undifferentiated programme. Identification of purely civilian facilities and programmes that have no strategic implications poses a particular challenge. Therefore, facilities identified as civilian in the Separation Plan will be offered for safeguards in phases to be decided by India. The nature of the facility concerned, the activities undertaken in it, the national security significance of materials and the location of the facilities are factors taken into account in undertaking the separation process. This is solely an Indian determination.

Three-Stage Programme

"7.The nuclear establishment in India not only built nuclear reactors but promoted the growth of a national industrial infrastructure. Nuclear power generation was envisaged as a three-stage programme with PHWRs (Pressurised Heavy Water Reactors) chosen for deployment in the first stage. As indigenous reactors were set up, several innovative design improvements were carried out based on Indian R&D and a standardized design was evolved. The research and technology development spanned the entire spectrum of the nuclear fuel cycle including the front end and the back end. Success in the technologies for the back end of the fuel cycle allowed us to launch the second stage of the programme by constructing a Fast Breeder Test Reactor. This reactor has operated for 20 years based on a unique carbide fuel and has achieved all technology objectives. We have now proceeded further and are constructing a 500 MWe Prototype Fast Breeder Reactor. Simultaneously, we have launched design and development of reactors aimed at thorium utilization and incorporating inherent safety features.

"8.Concepts such as grid connectivity are not relevant to the separation exercise. Issues related to fuel resource sustainability, technical design and economic viability, as well as smooth operation of reactors are relevant factors. This would necessitate grid connectivity irrespective of whether the reactor concerned is civilian or not civilian.

"9.It must be recognised that the Indian nuclear programme still has a relatively narrow base and cannot be expected to adopt solutions that might be deemed viable by much larger programmes. A comparison of the number of reactors and the total installed capacity between India and the P-5 brings this out graphically [see table].

"10. Another factor to be taken into account is the small capacity of the reactors produced indigenously by India, some of which would remain outside safeguards. Therefore, in assessing the extent of safeguards coverage, it would be important to look at both the number of reactors and the percentage of installed capacity covered. An average Indian reactor is of 220 MW and its output is significantly smaller than the standards reactor in a P-5 economy [see table].

"11. The complexity of the separation process is further enhanced by the limited resources that India has devoted to its nuclear programme as compared to P-5 nations. Moreover, as India expands international cooperation, the percentage of its thermal power reactor installed capacity under safeguards would rise significantly as fresh capacity is added through such cooperation.

"12. India's approach to the separation of its civilian nuclear facilities is guided by the following principles:

- Credible, feasible and implementable in a transparent manner;
- Consistent with the understandings of the 18 July Statement;
- Consistent with India's national security and R&D requirements as well as not prejudicial to the three-stage nuclear programme in India;
- Must be cost-effective in its implementation; and
- Must be acceptable to Parliament and public opinion.

"13. Based on these principles, India will:

- Include in the civilian list only those facilities offered for safeguards that, after separation, will no longer be engaged in activities of strategic significance.
- The overarching criterion would be a judgment whether subjecting a facility to IAEA safeguards would impact adversely on India's national security.
- However, a facility will be excluded from the civilian list if it is located in a larger hub of strategic significance, notwithstanding the fact that it may not be normally engaged in activities of strategic significance.
- A civilian facility would, therefore, be one that India has determined not to be relevant to its strategic programme.

"14. Taking the above into account, India, on the basis of reciprocal actions by the US, will adopt the following approach:

"(i) **Thermal Power Reactors:** India will identify and offer for safeguards 14 thermal power reactors between 2006 and 2014. This will include the 4 presently safeguarded reactors (TAPS 1&2, RAPS 1&2) and in addition KK 1&2 that are under construction. 8 other PHWRs, each of a capacity of 220 MW, will also be offered. Phasing of specific thermal power reactors, being offered for safeguards would be indicated separately by India. Such an offer would, in effect, cover 14 out of the 22 thermal power reactors in operation or currently under construction to be placed under safeguards, and would raise the total installed Thermal Power capacity by MWs under safeguards from the present 19% to 65% by 2014.

"(ii) **Fast Breeder Reactors:** India is not in a position to accept safeguards on the Prototype Fast Breeder Reactor (PFBR) and the Fast Breeder Test Reactor (FBTR), both located at Kalpakkam. The Fast Breeder Programme is at the R&D stage and its technology will take time to mature and reach an advanced stage of development.

"(iii) **Future Reactors:** India has decided to place under safeguards all future civilian thermal power reactors and civilian breeder reactors, and the Government of India retains the sole right to determine such reactors as civilian.

"(iv) **Research Reactors:** India will permanently shut down the CIRUS reactor, in 2010. It will also be prepared to shift the fuel core of the APSARA reactor that was purchased from France outside BARC (Bhabha Atomic Research Centre) and make the fuel core available to be placed under safeguards in 2010.

"(v) **Upstream facilities:** The following upstream facilities would be identified and separated as civilian:

- List of those specific facilities in the Nuclear Fuel Complex, which will be offered for safeguards by 2008 will be indicated separately.
- The Heavy Water Production plants at Thal, Tuticorin and Hazira are proposed to be designated for civilian use between 2006-2009. We do not consider these plants as relevant for safeguards purposes.

"(vi) **Downstream facilities:** The following downstream facilities would be identified and separated as civilian:

- India is willing to accept safeguards in the 'campaign' mode after 2010 in respect of the Tarapur Power Reactor Fuel Reprocessing Plant.
- The Tarapur and Rajasthan 'Away From Reactors' spent fuel storage pools would be made available for safeguards with appropriate phasing between 2006-2009.

"(vii) **Research Facilities:** India will declare the following facilities as civilian:

"(a) Tata Institute of Fundamental Research
"(b) Variable Energy Cyclotron Centre
"(c) Saha Institute of Nuclear Physics
"(d) Institute for Plasma Research
"(e) Institute of Mathematics Sciences
"(f) Institute of Physics
"(g) Tata Memorial Centre
"(h) Board of Radiation and Isotope Technology
"(i) Harish Chandra Research Institute

"These facilities are safeguards-irrelevant. It is our expectation that they will play a prominent role in international cooperation.

15. **Safeguards:**

"(a) The United States has conveyed its commitment to the reliable supply of fuel to India. Consistent with the July 18, 2005, Joint Statement, the United States has also reaffirmed its assurance to create the necessary conditions for India to have assured and full access to fuel for its reactors. As part of its implementation of the July 18, 2005, Joint Statement the United States is committed to seeking agreement from the U.S. Congress to amend its domestic laws and to work with friends and allies to adjust the practices of the Nuclear Suppliers Group to create the necessary conditions for India to obtain full access to the international fuel market, including reliable, uninterrupted and continual access to fuel supplies from firms in several nations.

"(b) To further guard against any disruption of fuel supplies, the United States is prepared to take the following additional steps:

"(i) The United States is willing to incorporate assurances regarding fuel supply in the bilateral US-India agreement on peaceful uses of nuclear energy under Section 123 of the US Atomic Energy Act, which would be submitted to the US Congress.

"(ii) The United States will join India in seeking to negotiate with the IAEA an India-specific fuel supply agreement.

"(iii) The United States will support an Indian effort to develop a strategic reserve of nuclear fuel to guard against any disruption of supply over the lifetime of India's reactors.

"(iv) If despite these arrangements, a disruption of fuel supplies to India occurs, the United States and India would jointly convene a group of friendly supplier countries to include countries such as Russia, France and the United Kingdom to pursue such measures as would restore fuel supply to India.

"(c) In light of the above understandings with the United States, an India-specific safeguards agreement will be negotiated between India and the IAEA providing for safeguards to guard against withdrawal of safeguarded nuclear material from civilian use at any time as well as providing for corrective measures that India may take to ensure uninterrupted operation of its civilian nuclear reactors in the event of disruption of foreign fuel supplies. Taking this into account, India will place its civilian nuclear facilities under India-specific safeguards in perpetuity and negotiate an appropriate safeguards agreement to this end with the IAEA.

"16. This plan is in conformity with the commitments made to Parliament by the Government."

Text of 123 Agreement

For Immediate Release August 3, 2007
2007/658

MEDIA NOTE
US and India Release Text of 123 Agreement

The following is the text of the Agreement for Cooperation between the Government of the United States of America and the Government of India concerning peaceful uses of nuclear energy (123 Agreement):

BEGIN TEXT:

AGREEMENT FOR COOPERATION BETWEEN THE GOVERNMENT OF THE UNITED STATES OF AMERICA AND THE GOVERNMENT OF INDIA CONCERNING PEACEFUL USES OF NUCLEAR ENERGY (123 AGREEMENT)

The Government of India and the Government of the United States of America, hereinafter referred to as the Parties,

RECOGNIZING the significance of civilian nuclear energy for meeting growing global energy demands in a cleaner and more efficient manner;

DESIRING to cooperate extensively in the full development and use of nuclear energy for peaceful purposes as a means of achieving energy security, on a stable, reliable and predictable basis;

WISHING to develop such cooperation on the basis of mutual respect for sovereignty, non-interference in each other's internal affairs, equality, mutual benefit, reciprocity and with due respect for each other's nuclear programmes;

DESIRING to establish the necessary legal framework and basis for cooperation concerning peaceful uses of nuclear energy;

AFFIRMING that cooperation under this Agreement is between two States possessing advanced nuclear technology, both Parties having the same benefits and advantages, both committed to preventing WMD proliferation;

NOTING the understandings expressed in the India-US Joint Statement of July 18, 2005 to enable full civil nuclear energy cooperation with India covering aspects of the associated nuclear fuel cycle;

AFFIRMING their support for the objectives of the International Atomic Energy Agency (IAEA) and its safeguards system, as applicable to India and the United States of America, and its importance in ensuring that international cooperation in development and use of nuclear energy for peaceful purposes is carried out under arrangements that will not contribute to the proliferation of nuclear weapons or other nuclear explosive devices;

NOTING their respective commitments to safety and security of peaceful uses of nuclear energy, to adequate physical protection of nuclear material and effective national export controls;

MINDFUL that peaceful nuclear activities must be undertaken with a view to protecting the environment;

MINDFUL of their shared commitment to preventing the proliferation of weapons of mass destruction; and

DESIROUS of strengthening the strategic partnership between them;

Have agreed on the following:

ARTICLE 1 – DEFINITIONS

For the purposes of this Agreement:

(A) "By-product material" means any radioactive material (except special fissionable material) yielded in or made radioactive by exposure to the radiation incident to the process of producing or utilizing special fissionable material. By-product material shall not be subject to safeguards or any other form of verification under this Agreement, unless it has been decided otherwise by prior mutual agreement in writing between the two Parties.

(B) "Component" means a component part of equipment, or other item so designated by agreement of the Parties.

(C) "Conversion" means any of the normal operations in the nuclear fuel cycle, preceding fuel fabrication and excluding enrichment, by which uranium is transformed from one chemical form to another – for example, from uranium hexafluoride (UF6) to uranium dioxide (UO2) or from uranium oxide to metal.

(D) "Decommissioning" means the actions taken at the end of a facility's useful life to retire the facility from service in the manner that provides adequate protection for the health and safety of the decommissioning workers and the general public, and for the environment. These actions can range from closing down the facility and a minimal removal of nuclear material coupled with continuing maintenance and surveillance, to complete removal of residual radioactivity in excess of levels acceptable for unrestricted use of the facility and its site.

(E) “Dual-Use Item” means a nuclear related item which has a technical use in both nuclear and non-nuclear applications.

(F) “Equipment” means any equipment in nuclear operation including reactor, reactor pressure vessel, reactor fuel charging and discharging equipment, reactor control rods, reactor pressure tubes, reactor primary coolant pumps, zirconium tubing, equipment for fuel fabrication and any other item so designated by the Parties.

(G) “High enriched uranium” means uranium enriched to twenty percent or greater in the isotope 235.

(H) “Information” means any information that is not in the public domain and is transferred in any form pursuant to this Agreement and so designated and documented in hard copy or digital form by mutual agreement by the Parties that it shall be subject to this Agreement, but will cease to be information whenever the Party transferring the information or any third party legitimately releases it into the public domain.

(I) “Low enriched uranium” means uranium enriched to less than twenty percent in the isotope 235.

(J) “Major critical component” means any part or group of parts essential to the operation of a sensitive nuclear facility or heavy water production facility.

(K) “Non-nuclear material” means heavy water, or any other material suitable for use in a reactor to slow down high velocity neutrons and increase the likelihood of further fission, as may be jointly designated by the appropriate authorities of the Parties.

(L) “Nuclear material” means (1) source material and (2) special fissionable material. “Source material” means uranium containing the mixture of isotopes occurring in nature; uranium depleted in the isotope 235; thorium; any of the foregoing in the form of metal, alloy, chemical compound, or concentrate; any other material containing one or more of the foregoing in such concentration as the Board of Governors of the

IAEA shall from time to time determine; and such other materials as the Board of Governors of the IAEA may determine or as may be agreed by the appropriate authorities of both Parties. "Special fissionable material" means plutonium, uranium-233, uranium enriched in the isotope 233 or 235, any substance containing one or more of the foregoing, and such other substances as the Board of Governors of the IAEA may determine or as may be agreed by the appropriate authorities of both Parties. "Special fissionable material" does not include "source material". Any determination by the Board of Governors of the IAEA under Article XX of that Agency's Statute or otherwise that amends the list of materials considered to be "source material" or "special fissionable material" shall only have effect under this Agreement when both Parties to this Agreement have informed each other in writing that they accept such amendment.

(M) "Peaceful purposes" include the use of information, nuclear material, equipment or components in such fields as research, power generation, medicine, agriculture and industry, but do not include use in, research on, or development of any nuclear explosive device or any other military purpose. Provision of power for a military base drawn from any power network, production of radioisotopes to be used for medical purposes in military environment for diagnostics, therapy and sterility assurance, and other similar purposes as may be mutually agreed by the Parties shall not be regarded as military purpose.

(N) "Person" means any individual or any entity subject to the territorial jurisdiction of either Party but does not include the Parties.

(O) "Reactor" means any apparatus, other than a nuclear weapon or other nuclear explosive device, in which a self-sustaining fission chain reaction is maintained by utilizing uranium, plutonium, or thorium or any combination thereof.

(P) "Sensitive nuclear facility" means any facility designed or used primarily for uranium enrichment, reprocessing of nuclear fuel, or fabrication of nuclear fuel containing plutonium.

(Q) "Sensitive nuclear technology" means any information that is not in the public domain and that is important to the design, construction, fabrication, operation, or maintenance of any sensitive nuclear facility, or other such information that may be so designated by agreement of the Parties.

ARTICLE 2 – SCOPE OF COOPERATION

1. The Parties shall cooperate in the use of nuclear energy for peaceful purposes in accordance with the provisions of this Agreement. Each Party shall implement this Agreement in accordance with its respective applicable treaties, national laws, regulations, and license requirements concerning the use of nuclear energy for peaceful purposes.

2. The purpose of the Agreement being to enable full civil nuclear energy cooperation between the Parties, the Parties may pursue cooperation in all relevant areas to include, but not limited to, the following:

a. Advanced nuclear energy research and development in such areas as may be agreed between the Parties;
b. Nuclear safety matters of mutual interest and competence, as set out in Article 3;
c. Facilitation of exchange of scientists for visits, meetings, symposia and collaborative research;
d. Full civil nuclear cooperation activities covering nuclear reactors and aspects of the associated nuclear fuel cycle including technology transfer on an industrial or commercial scale between the Parties or authorized persons;
e. Development of a strategic reserve of nuclear fuel to guard against any disruption of supply over the lifetime of India's reactors;
f. Advanced research and development in nuclear sciences including but not limited to biological research, medicine, agriculture and industry, environment and climate change;
g. Supply between the Parties, whether for use by or for the benefit of the Parties or third countries, of nuclear material;
h. Alteration in form or content of nuclear material as provided for in Article 6;
i. Supply between the Parties of equipment, whether for use by or for the benefit of the Parties or third countries;
j. Controlled thermonuclear fusion including in multilateral projects; and
k. Other areas of mutual interest as may be agreed by the Parties.

3. Transfer of nuclear material, non-nuclear material, equipment, components and information under this Agreement may be undertaken

directly between the Parties or through authorized persons. Such transfers shall be subject to this Agreement and to such additional terms and conditions as may be agreed by the Parties. Nuclear material, non-nuclear material, equipment, components and information transferred from the territory of one Party to the territory of the other Party, whether directly or through a third country, will be regarded as having been transferred pursuant to this Agreement only upon confirmation, by the appropriate authority of the recipient Party to the appropriate authority of the supplier Party that such items both will be subject to the Agreement and have been received by the recipient Party.

4. The Parties affirm that the purpose of this Agreement is to provide for peaceful nuclear cooperation and not to affect the unsafeguarded nuclear activities of either Party. Accordingly, nothing in this Agreement shall be interpreted as affecting the rights of the Parties to use for their own purposes nuclear material, non-nuclear material, equipment, components, information or technology produced, acquired or developed by them independent of any nuclear material, non-nuclear material, equipment, components, information or technology transferred to them pursuant to this Agreement. This Agreement shall be implemented in a manner so as not to hinder or otherwise interfere with any other activities involving the use of nuclear material, non-nuclear material, equipment, components, information or technology and military nuclear facilities produced, acquired or developed by them independent of this Agreement for their own purposes.

ARTICLE 3 – TRANSFER OF INFORMATION

1. Information concerning the use of nuclear energy for peaceful purposes may be transferred between the Parties. Transfers of information may be accomplished through reports, data banks and computer programs and any other means mutually agreed to by the Parties. Fields that may be covered include, but shall not be limited to, the following:

a. Research, development, design, construction, operation, maintenance and use of reactors, reactor experiments, and decommissioning;
b. The use of nuclear material in physical, chemical, radiological and biological research, medicine, agriculture and industry;
c. Fuel cycle activities to meet future world-wide civil nuclear energy needs, including multilateral approaches to which they are parties

for ensuring nuclear fuel supply and appropriate techniques for management of nuclear wastes;

d. Advanced research and development in nuclear science and technology;
e. Health, safety, and environmental considerations related to the foregoing;
f. Assessments of the role nuclear power may play in national energy plans;
g. Codes, regulations and standards for the nuclear industry;
h. Research on controlled thermonuclear fusion including bilateral activities and contributions toward multilateral projects such as the International Thermonuclear Experimental Reactor (ITER); and
i. Any other field mutually agreed to by the Parties.

2. Cooperation pursuant to this Article may include, but is not limited to, training, exchange of personnel, meetings, exchange of samples, materials and instruments for experimental purposes and a balanced participation in joint studies and projects.

3. This Agreement does not require the transfer of any information regarding matters outside the scope of this Agreement, or information that the Parties are not permitted under their respective treaties, national laws, or regulations to transfer.

4. Restricted Data, as defined by each Party, shall not be transferred under this Agreement.

ARTICLE 4 – NUCLEAR TRADE

1. The Parties shall facilitate nuclear trade between themselves in the mutual interests of their respective industry, utilities and consumers and also, where appropriate, trade between third countries and either Party of items obligated to the other Party. The Parties recognize that reliability of supplies is essential to ensure smooth and uninterrupted operation of nuclear facilities and that industry in both the Parties needs continuing reassurance that deliveries can be made on time in order to plan for the efficient operation of nuclear installations.

2. Authorizations, including export and import licenses as well as authorizations or consents to third parties, relating to trade, industrial operations or nuclear material movement should be consistent with the

sound and efficient administration of this Agreement and should not be used to restrict trade. It is further agreed that if the relevant authority of the concerned Party considers that an application cannot be processed within a two month period it shall immediately, upon request, provide reasoned information to the submitting Party. In the event of a refusal to authorize an application or a delay exceeding four months from the date of the first application the Party of the submitting persons or undertakings may call for urgent consultations under Article 13 of this Agreement, which shall take place at the earliest opportunity and in any case not later than 30 days after such a request.

ARTICLE 5 – TRANSFER OF NUCLEAR MATERIAL, NON-NUCLEAR MATERIAL, EQUIPMENT, COMPONENTS AND RELATED TECHNOLOGY

1. Nuclear material, non-nuclear material, equipment and components may be transferred for applications consistent with this Agreement. Any special fissionable material transferred under this Agreement shall be low enriched uranium, except as provided in paragraph 5.

2. Sensitive nuclear technology, heavy water production technology, sensitive nuclear facilities, heavy water production facilities and major critical components of such facilities may be transferred under this Agreement pursuant to an amendment to this Agreement. Transfers of dual-use items that could be used in enrichment, reprocessing or heavy water production facilities will be subject to the Parties' respective applicable laws, regulations and license policies.

3. Natural or low enriched uranium may be transferred for use as fuel in reactor experiments and in reactors, for conversion or fabrication, or for such other purposes as may be agreed to by the Parties.

4. The quantity of nuclear material transferred under this Agreement shall be consistent with any of the following purposes: use in reactor experiments or the loading of reactors, the efficient and continuous conduct of such reactor experiments or operation of reactors for their lifetime, use as samples, standards, detectors, and targets, and the accomplishment of other purposes as may be agreed by the Parties.

5. Small quantities of special fissionable material may be transferred for use as samples, standards, detectors, and targets, and for such other

purposes as the Parties may agree.

6. (a) The United States has conveyed its commitment to the reliable supply of fuel to India. Consistent with the July 18, 2005, Joint Statement, the United States has also reaffirmed its assurance to create the necessary conditions for India to have assured and full access to fuel for its reactors. As part of its implementation of the July 18, 2005, Joint Statement the United States is committed to seeking agreement from the US Congress to amend its domestic laws and to work with friends and allies to adjust the practices of the Nuclear Suppliers Group to create the necessary conditions for India to obtain full access to the international fuel market, including reliable, uninterrupted and continual access to fuel supplies from firms in several nations.

(b) To further guard against any disruption of fuel supplies, the United States is prepared to take the following additional steps:

i) The United States is willing to incorporate assurances regarding fuel supply in the bilateral US-India agreement on peaceful uses of nuclear energy under Section 123 of the US Atomic Energy Act, which would be submitted to the US Congress.

ii) The United States will join India in seeking to negotiate with the IAEA an India-specific fuel supply agreement.

iii) The United States will support an Indian effort to develop a strategic reserve of nuclear fuel to guard against any disruption of supply over the lifetime of India's reactors.

iv) If despite these arrangements, a disruption of fuel supplies to India occurs, the United States and India would jointly convene a group of friendly supplier countries to include countries such as Russia, France and the United Kingdom to pursue such measures as would restore fuel supply to India.

(c) In light of the above understandings with the United States, an India-specific safeguards agreement will be negotiated between India and the IAEA providing for safeguards to guard against withdrawal of safeguarded nuclear material from civilian use at any time as well as providing for corrective measures that India may take to ensure uninterrupted operation of its civilian nuclear reactors in the event of disruption of foreign fuel supplies. Taking this into account, India will place its civilian nuclear facilities under India-specific safeguards in perpetuity and negotiate an appropriate safeguards agreement to this end with the IAEA.

ARTICLE 6 – NUCLEAR FUEL CYCLE ACTIVITIES

In keeping with their commitment to full civil nuclear cooperation, both Parties, as they do with other states with advanced nuclear technology, may carry out the following nuclear fuel cycle activities:

i) Within the territorial jurisdiction of either Party, enrichment up to twenty percent in the isotope 235 of uranium transferred pursuant to this Agreement, as well as of uranium used in or produced through the use of equipment so transferred, may be carried out.

ii) Irradiation within the territorial jurisdiction of either Party of plutonium, uranium-233, high enriched uranium and irradiated nuclear material transferred pursuant to this Agreement or used in or produced through the use of non-nuclear material, nuclear material or equipment so transferred may be carried out.

iii) With a view to implementing full civil nuclear cooperation as envisioned in the Joint Statement of the Parties of July 18, 2005, the Parties grant each other consent to reprocess or otherwise alter in form or content nuclear material transferred pursuant to this Agreement and nuclear material and by-product material used in or produced through the use of nuclear material, non-nuclear material, or equipment so transferred. To bring these rights into effect, India will establish a new national reprocessing facility dedicated to reprocessing safeguarded nuclear material under IAEA safeguards and the Parties will agree on arrangements and procedures under which such reprocessing or other alteration in form or content will take place in this new facility. Consultations on arrangements and procedures will begin within six months of a request by either Party and will be concluded within one year. The Parties agree on the application of IAEA safeguards to all facilities concerned with the above activities. These arrangements and procedures shall include provisions with respect to physical protection standards set out in Article 8, storage standards set out in Article 7, and environmental protections set forth in Article 11 of this Agreement, and such other provisions as may be agreed by the Parties. Any special fissionable material that may be separated may only be utilized in national facilities under IAEA safeguards.

iv) Post-irradiation examination involving chemical dissolution or separation of irradiated nuclear material transferred pursuant to this Agreement or irradiated nuclear material used in or produced through the use of non-nuclear material, nuclear material or equipment so transferred may be carried out.

ARTICLE 7 – STORAGE AND RETRANSFERS

1. Plutonium and uranium 233 (except as either may be contained in irradiated fuel elements), and high enriched uranium, transferred pursuant to this Agreement or used in or produced through the use of material or equipment so transferred, may be stored in facilities that are at all times subject, as a minimum, to the levels of physical protection that are set out in IAEA document INFCIRC 225/REV 4 as it may be revised and accepted by the Parties. Each Party shall record such facilities on a list, made available to the other Party. A Party's list shall be held confidential if that Party so requests. Either Party may make changes to its list by notifying the other Party in writing and receiving a written acknowledgement. Such acknowledgement shall be given no later than thirty days after the receipt of the notification and shall be limited to a statement that the notification has been received. If there are grounds to believe that the provisions of this sub-Article are not being fully complied with, immediate consultations may be called for. Following upon such consultations, each Party shall ensure by means of such consultations that necessary remedial measures are taken immediately. Such measures shall be sufficient to restore the levels of physical protection referred to above at the facility in question. However, if the Party on whose territory the nuclear material in question is stored determines that such measures are not feasible, it will shift the nuclear material to another appropriate, listed facility it identifies.

2. Nuclear material, non-nuclear material, equipment, components, and information transferred pursuant to this Agreement and any special fissionable material produced through the use of nuclear material, non-nuclear material or equipment so transferred shall not be transferred or re-transferred to unauthorized persons or, unless the Parties agree, beyond the recipient Party's territorial jurisdiction.

ARTICLE 8 – PHYSICAL PROTECTION

1. Adequate physical protection shall be maintained with respect to

nuclear material and equipment transferred pursuant to this Agreement and nuclear material used in or produced through the use of nuclear material, non-nuclear material or equipment so transferred.

2. To fulfill the requirement in paragraph 1, each Party shall apply measures in accordance with (i) levels of physical protection at least equivalent to the recommendations published in IAEA document INFCIRC/225/Rev.4 entitled "The Physical Protection of Nuclear Material and Nuclear Facilities," and in any subsequent revisions of that document agreed to by the Parties, and (ii) the provisions of the 1980 Convention on the Physical Protection of Nuclear Material and any amendments to the Convention that enter into force for both Parties.

3. The Parties will keep each other informed through diplomatic channels of those agencies or authorities having responsibility for ensuring that levels of physical protection for nuclear material in their territory or under their jurisdiction or control are adequately met and having responsibility for coordinating response and recovery operations in the event of unauthorized use or handling of material subject to this Article. The Parties will also keep each other informed through diplomatic channels of the designated points of contact within their national authorities to cooperate on matters of out-of-country transportation and other matters of mutual concern.

4. The provisions of this Article shall be implemented in such a manner as to avoid undue interference in the Parties' peaceful nuclear activities and so as to be consistent with prudent management practices required for the safe and economic conduct of their peaceful nuclear programs.

ARTICLE 9 – PEACEFUL USE

Nuclear material, equipment and components transferred pursuant to this Agreement and nuclear material and by-product material used in or produced through the use of any nuclear material, equipment, and components so transferred shall not be used by the recipient Party for any nuclear explosive device, for research on or development of any nuclear explosive device or for any military purpose.

ARTICLE 10 – IAEA SAFEGUARDS

1. Safeguards will be maintained with respect to all nuclear materials and equipment transferred pursuant to this Agreement, and with

respect to all special fissionable material used in or produced through the use of such nuclear materials and equipment, so long as the material or equipment remains under the jurisdiction or control of the cooperating Party.

2. Taking into account Article 5.6 of this Agreement, India agrees that nuclear material and equipment transferred to India by the United States of America pursuant to this Agreement and any nuclear material used in or produced through the use of nuclear material, non-nuclear material, equipment or components so transferred shall be subject to safeguards in perpetuity in accordance with the India-specific Safeguards Agreement between India and the IAEA (identifying data) and an Additional Protocol, when in force.

3. Nuclear material and equipment transferred to the United States of America pursuant to this Agreement and any nuclear material used in or produced through the use of any nuclear material, non-nuclear material, equipment, or components so transferred shall be subject to the Agreement between the United States of America and the IAEA for the application of safeguards in the United States of America, done at Vienna November 18, 1977, which entered into force on December 9, 1980, and an Additional Protocol, when in force.

4. If the IAEA decides that the application of IAEA safeguards is no longer possible, the supplier and recipient should consult and agree on appropriate verification measures.

5. Each Party shall take such measures as are necessary to maintain and facilitate the application of IAEA safeguards in its respective territory provided for under this Article.

6. Each Party shall establish and maintain a system of accounting for and control of nuclear material transferred pursuant to this Agreement and nuclear material used in or produced through the use of any material, equipment, or components so transferred. The procedures applicable to India shall be those set forth in the India-specific Safeguards Agreement referred to in Paragraph 2 of this Article.

7. Upon the request of either Party, the other Party shall report or permit the IAEA to report to the requesting Party on the status of all inventories of material subject to this Agreement.

8. The provisions of this Article shall be implemented in such a manner as to avoid hampering, delay, or undue interference in the

Parties' peaceful nuclear activities and so as to be consistent with prudent management practices required for the safe and economic conduct of their peaceful nuclear programs.

ARTICLE 11 – ENVIRONMENTAL PROTECTION

The Parties shall cooperate in following the best practices for minimizing the impact on the environment from any radioactive, chemical or thermal contamination arising from peaceful nuclear activities under this Agreement and in related matters of health and safety.

ARTICLE 12 – IMPLEMENTATION OF THE AGREEMENT

1. This Agreement shall be implemented in a manner designed:

a) to avoid hampering or delaying the nuclear activities in the territory of either Party;
b) to avoid interference in such activities;
c) to be consistent with prudent management practices required for the safe conduct of such activities; and
d) to take full account of the long term requirements of the nuclear energy programs of the Parties.

2. The provisions of this Agreement shall not be used to:

a) secure unfair commercial or industrial advantages or to restrict trade to the disadvantage of persons and undertakings of either Party or hamper their commercial or industrial interests, whether international or domestic;
b) interfere with the nuclear policy or programs for the promotion of the peaceful uses of nuclear energy including research and development; or
c) impede the free movement of nuclear material, non nuclear material and equipment supplied under this Agreement within the territory of the Parties.

3. When execution of an agreement or contract pursuant to this Agreement between Indian and United States organizations requires exchanges of experts, the Parties shall facilitate entry of the experts to their territories and their stay therein consistent with national laws, regulations and practices. When other cooperation pursuant to this Agreement requires visits of experts, the Parties shall facilitate entry

of the experts to their territory and their stay therein consistent with national laws, regulations and practices.

ARTICLE 13 – CONSULTATIONS

1. The Parties undertake to consult at the request of either Party regarding the implementation of this Agreement and the development of further cooperation in the field of peaceful uses of nuclear energy on a stable, reliable and predictable basis. The Parties recognize that such consultations are between two States with advanced nuclear technology, which have agreed to assume the same responsibilities and practices and acquire the same benefits and advantages as other leading countries with advanced nuclear technology.

2. Each Party shall endeavor to avoid taking any action that adversely affects cooperation envisaged under Article 2 of this Agreement. If either Party at any time following the entry into force of this Agreement does not comply with the provisions of this Agreement, the Parties shall promptly hold consultations with a view to resolving the matter in a way that protects the legitimate interests of both Parties, it being understood that rights of either Party under Article 16.2 remain unaffected.

3. Consultations under this Article may be carried out by a Joint Committee specifically established for this purpose. A Joint Technical Working Group reporting to the Joint Committee will be set up to ensure the fulfillment of the requirements of the Administrative Arrangements referred to in Article 17.

ARTICLE 14 – TERMINATION AND CESSATION OF COOPERATION

1. Either Party shall have the right to terminate this Agreement prior to its expiration on one year's written notice to the other Party. A Party giving notice of termination shall provide the reasons for seeking such termination. The Agreement shall terminate one year from the date of the written notice, unless the notice has been withdrawn by the providing Party in writing prior to the date of termination.

2. Before this Agreement is terminated pursuant to paragraph 1 of this Article, the Parties shall consider the relevant circumstances and promptly hold consultations, as provided in Article 13, to address the reasons cited by the Party seeking termination. The Party seeking termination has the

right to cease further cooperation under this Agreement if it determines that a mutually acceptable resolution of outstanding issues has not been possible or cannot be achieved through consultations. The Parties agree to consider carefully the circumstances that may lead to termination or cessation of cooperation. They further agree to take into account whether the circumstances that may lead to termination or cessation resulted from a Party's serious concern about a changed security environment or as a response to similar actions by other States which could impact national security.

3. If a Party seeking termination cites a violation of this Agreement as the reason for notice for seeking termination, the Parties shall consider whether the action was caused inadvertently or otherwise and whether the violation could be considered as material. No violation may be considered as being material unless corresponding to the definition of material violation or breach in the Vienna Convention on the Law of Treaties. If a Party seeking termination cites a violation of an IAEA safeguards agreement as the reason for notice for seeking termination, a crucial factor will be whether the IAEA Board of Governors has made a finding of non-compliance.

4. Following the cessation of cooperation under this Agreement, either Party shall have the right to require the return by the other Party of any nuclear material, equipment, non-nuclear material or components transferred under this Agreement and any special fissionable material produced through their use. A notice by a Party that is invoking the right of return shall be delivered to the other Party on or before the date of termination of this Agreement. The notice shall contain a statement of the items subject to this Agreement as to which the Party is requesting return. Except as provided in provisions of Article 16.3, all other legal obligations pertaining to this Agreement shall cease to apply with respect to the nuclear items remaining on the territory of the Party concerned upon termination of this Agreement.

5. The two Parties recognize that exercizing the right of return would have profound implications for their relations. If either Party seeks to exercize its right pursuant to paragraph 4 of this Article, it shall, prior to the removal from the territory or from the control of the other Party of any nuclear items mentioned in paragraph 4,

undertake consultations with the other Party. Such consultations shall give special consideration to the importance of uninterrupted operation of nuclear reactors of the Party concerned with respect to the availability of nuclear energy for peaceful purposes as a means of achieving energy security. Both Parties shall take into account the potential negative consequences of such termination on the on-going contracts and projects initiated under this Agreement of significance for the respective nuclear programmes of either Party.

6. If either Party exercizes its right of return pursuant to paragraph 4 of this Article, it shall, prior to the removal from the territory or from the control of the other Party, compensate promptly that Party for the fair market value thereof and for the costs incurred as a consequence of such removal. If the return of nuclear items is required, the Parties shall agree on methods and arrangements for the return of the items, the relevant quantity of the items to be returned, and the amount of compensation that would have to be paid by the Party exercizing the right to the other Party.

7. Prior to return of nuclear items, the Parties shall satisfy themselves that full safety, radiological and physical protection measures have been ensured in accordance with their existing national regulations and that the transfers pose no unreasonable risk to either Party, countries through which the nuclear items may transit and to the global environment and are in accordance with existing international regulations.

8. The Party seeking the return of nuclear items shall ensure that the timing, methods and arrangements for return of nuclear items are in accordance with paragraphs 5, 6 and 7. Accordingly, the consultations between the Parties shall address mutual commitments as contained in Article 5.6. It is not the purpose of the provisions of this Article regarding cessation of cooperation and right of return to derogate from the rights of the Parties under Article 5.6.

9. The arrangements and procedures concluded pursuant to Article 6(iii) shall be subject to suspension by either Party in exceptional circumstances, as defined by the Parties, after consultations have been held between the Parties aimed at reaching mutually acceptable resolution of outstanding issues, while taking into account the effects of such suspension on other aspects of cooperation under this Agreement.

ARTICLE 15 – SETTLEMENT OF DISPUTES

Any dispute concerning the interpretation or implementation of the provisions of this Agreement shall be promptly negotiated by the Parties with a view to resolving that dispute.

ARTICLE 16 – ENTRY INTO FORCE AND DURATION

1. This Agreement shall enter into force on the date on which the Parties exchange diplomatic notes informing each other that they have completed all applicable requirements for its entry into force.

2. This Agreement shall remain in force for a period of 40 years. It shall continue in force thereafter for additional periods of 10 years each. Each Party may, by giving 6 months written notice to the other Party, terminate this Agreement at the end of the initial 40 year period or at the end of any subsequent 10 year period.

3. Notwithstanding the termination or expiration of this Agreement or withdrawal of a Party from this Agreement, Articles 5.6(c), 6, 7, 8, 9, 10 and 15 shall continue in effect so long as any nuclear material, non-nuclear material, by-product material, equipment or components subject to these articles remains in the territory of the Party concerned or under its jurisdiction or control anywhere, or until such time as the Parties agree that such nuclear material is no longer usable for any nuclear activity relevant from the point of view of safeguards.

4. This Agreement shall be implemented in good faith and in accordance with the principles of international law.

5. The Parties may consult, at the request of either Party, on possible amendments to this Agreement. This Agreement may be amended if the Parties so agree. Any amendment shall enter into force on the date on which the Parties exchange diplomatic notes informing each other that their respective internal legal procedures necessary for the entry into force have been completed.

ARTICLE 17 – ADMINISTRATIVE ARRANGEMENT

1. The appropriate authorities of the Parties shall establish an Administrative Arrangement in order to provide for the effective implementation of the provisions of this Agreement.

2. The principles of fungibility and equivalence shall apply to nuclear material and non-nuclear material subject to this Agreement.

Detailed provisions for applying these principles shall be set forth in the Administrative Arrangement.

3. The Administrative Arrangement established pursuant to this Article may be amended by agreement of the appropriate authorities of the Parties.

IN WITNESS WHEREOF the undersigned, being duly authorized, have signed this Agreement.

DONE at, this day of, 200 , in duplicate.

FOR THE GOVERNMENT OF THE UNITED STATES OF AMERICA:	**FOR THE GOVERNMENT OF INDIA:**

AGREED MINUTE

During the negotiation of the Agreement for Cooperation Between the Government of the United States of America and the Government of India Concerning Peaceful Uses of Nuclear Energy ("the Agreement") signed today, the following understandings, which shall be an integral part of the Agreement, were reached.

Proportionality

For the purposes of implementing the rights specified in Articles 6 and 7 of the Agreement with respect to special fissionable material and by-product material produced through the use of nuclear material and non-nuclear material, respectively, transferred pursuant to the Agreement and not used in or produced through the use of equipment transferred pursuant to the Agreement, such rights shall in practice be applied to that proportion of special fissionable material and by-product material produced that represents the ratio of transferred nuclear material and non-nuclear material, respectively, used in the production of the special fissionable material and by-product material to the total amount of nuclear material and non-nuclear material so used, and similarly for subsequent generations.

By-product material

The Parties agree that reporting and exchanges of information on by-product material subject to the Agreement will be limited to the following:

(1) Both Parties would comply with the provisions as contained in the IAEA document GOV/1999/19/Rev.2, with regard to by-product material subject to the Agreement.

(2) With regard to tritium subject to the Agreement, the Parties will exchange annually information pertaining to its disposition for peaceful purposes consistent with Article 9 of this Agreement.

FOR THE GOVERNMENT OF THE UNITED STATES OF AMERICA:	FOR THE GOVERNMENT OF INDIA:

Appendix 6

Action Plan for Ushering in a Nuclear Weapon-Free and Non-Violent World Order

1. Humanity stands at a crossroads of history. The world has lived too long under the sentence of extinction. Nuclear weapons threaten to annihilate human civilization and all that humankind has built thought millennia of labour and toil. Nuclear-weapon States and non-nuclear – weapon States alike are threatened by such a holocaust. It is imperative that nuclear weapons be eliminated. The recently signed INF Treaty between the United States and the Soviet Union is a first major step in this direction. This process must be taken to its logical conclusion by ridding the world of nuclear weapons. The time has also come to consider seriously the changes in doctrines, in policies, in attitudes, and in the institutions required to usher in and manage a nuclear weapon-free and non-violent world. Peace must be predicated on a basis other than the assurance of global destruction. We need a world order based on non-violence and peaceful coexistence. We need international institutions that will nurture such a world order.

2. We call upon the international community to urgently negotiate a binding commitment to an action plan for ushering in a non-violent world free of nuclear weapons. We suggest the following action plan as a basis for such negotiations.

1.1. STAGE 1 (duration: 6 years, from 1988 to 194)

2.1.a. *Nuclear disarmament:*

2.1.a.i. Elimination of all Soviet and United States land-based medium and shorter-range missiles (500 to 5,500 kilometers) in accordance with the INF Treaty.

2.1.a.ii. Agreement on a 50 percent cut in Soviet and United States strategic arsenals (with ranges above 5,500 kilometers)

2.1.a.iii. Agreement on a phased elimination by the year 2000 A.D. of United States and Soviet short range battlefield and air-launched nuclear weapons.

2.1.a.iv. Cessation of the production of nuclear-weapons by all nuclear-weapon States.

2.1.a.v. Cessation of production of weapon grade fissionable material by all nuclear weapon States.

2.1.a.vi. Moratorium on the testing of nuclear weapons.

2.1.a.vii. Commencement ad conclusion of negotiations on a comprehensive test ban treaty.

2.1.b. *Measures collateral to nuclear disarmament*:

2.1.b.i. Conclusion of a convention to outlaw the use and threat of use of nuclear weapons pending their elimination.

2.1.b.ii. Declaration by the United States and the Soviet Union that the fissile materal released under the INF Treaty would be utilized for peaceful purposes only and accordingly be subjected to supervision by the International Atomic Energy Agency.

2.1.b.iii. Declaration by all nuclear weapon States of their stockpiles of nuclear weapons and weapon-grade fissionable material.

2.1.b.iv. Cessation of direct of indirect transfer to other States of nuclear weapons, delivery systems, and weapon-grade fissionable material.

2.1.b.v. Non-nuclear weapon Powers to undertake not to cross the threshold into the acquisition of nuclear weapons.

2.1.b.vi. Initiation of multilateral negotiations to be concluded by 1995, for weapons by the year 2010. This treaty would replace the Non-Proliferation Treaty, which ends it 1995.

2.1.b.v. Non-nuclear weapon Powers to undertake not to cross the threshold into the acquisition of nuclear weapons.

2.1.b.vi. Initiation of multilateral negotiations to be concluded by 1995, for a new treaty eliminating all nuclear weapons by the year 2010. This treaty would replace the Non-Proliferation Treaty, which ends it 1995.

2.1.c. *Other weapons of mass destruction.*

2.1.c.i. Conclusion of a treaty banning chemical weapons

2.1.c.ii. Conclusion of a treaty banning radio logical weapons.

2.1.d. *Conventional force*:

2.1.d.i. Substantial reduction of NATO and Warsaw Pact conventional force especially offensive forces, and weapon systems in Europe from the Atlantic to the Urals.

2.1.d.ii. Multilateral discussions in the Conference on Disarmament or in the United Nations on military doctrines with a view to working towards the goal of a purely defensive orientation for the armed forced of the world. The discussion would include measures to prevent surprise attacks.

2.1.e. *Space weapon systems:*

2.1.e.i. A moratorium on the testing and deployment of all space weapon systems.

2.1.e.ii. Expansion of international cooperation in the peaceful uses of outer space.

2.1.f. *Control and management of the arms race based on new technologies.*

2.1.f.i. Arrangements for monitoring and assessing new technologies which have military applications as well as forecasting their implications for international security.

2.1.f.ii. For research in frontier areas of technology where there are potential military applications, new technology projects and technological missions should be undertaken under the auspices of the United Nations in order to direct them exclusively to civilian sectors.

2.1.f.iii. Commencement of work, under the aegis of the United Nations, for the formulation of guidelines to be observed by Governments in respect of new technologies with potential military applications.

2.1.f.iv. Commencement of negotiations for banning technological missions designed to develop new weapon systems and means of warfare.

2.1.g *Verification:*

2.1.g.i. Acceptance in principle of the need to establish an integrated multilateral verification system under the aegis of the United Nations as an integral part of a strengthened multilateral framework required to ensure peace and security during the process of disarmament as well as in a nuclear-weapon-free world.

2.2. STAGE II (duration: 6 years, from 1995 to 2000)

2.2.a.i. Nuclear disarmament:

2.2.a.i Completion of Stage I reductions by the United States and the Soviet Union and the induction of all other nuclear weapon States into the process of nuclear disarmament.

2.2.a.ii. Elimination of all medium-and short-range, sea-based, land-based and air-launched nuclear missilies by all nuclear weapon States.

2.2.a.iii. Elimination of all tactical battlefield nuclear weapons)land, sea and air) by all nuclear weapon States.

2.2.a.iv. Entry into force of the Comprehensive Test Ban Treaty.

2.2.b. *Measures collateral to nuclear disarmament:*

2.2.b.i. Negotiations on the withdrawal of strategic nuclear weapons deployed beyond national boundaries.

2.2.b.ii. Completion of the ratification and entry into force of the convention prohibiting the use and threat of use of nuclear weapons.

2.2.b.iii. Conclusion of the new treaty elimination all nuclear weapons by the year 2010 to replace the Non-Proliferation Treaty.

2.2.c. *Space weapons:*

2.2.c.i. Agreement within a multilateral framework on banning the testing development, deployment and strorage of all space weapons.

2.2.d. *Conventional forces:*

2.2.d.i. Further reduction of NATO and Warsaw Pact conventional forces to minimum defensive levels.

2.2.d.ii. Negotiations under the Conference on Disarmament on global conventional arms reduction.

2.2.d.iii. Removal of all military forces and bases from foreign territories.

2.2.e. New and emerging technologies:

2.2.e.i. Completion of negotiations on banning technological missions aimed at the development of new weapon systems.

2.2.e.ii. Completion of negotiations on guidelines in respect of new technologies with potential military applications

2.2.f. *Comprehensive global security system:*

2.2.f.i. Negotiations on and establishment of a comprehensive global security system to sustain a world without nuclear weapons. This would include institutional steps to ensure the effective implementation of the provisions of the Charter of the United Nations relating to the non-use of force, the peaceful settlement of disputes, and the right of every State to pursue its own path of development.

2.2.f.ii. Arrangements for the release of resources through disarmament for development purposes.

2.2.f.iii. Elimination of non-military threats to security by such measures as the establishment of a just and equitable international economic order.

2.2.f.iv. The strengthening of United Nations system and related multilateral forums.

2.2.f.v. The commencement of negotiations for the establishment of an integrated multilateral verification system under the United Nations.

2.3. STAGE III (duration: 10 years, from 2001 to 2010)

2.3.a. Elimination of all nuclear weapons from the world.

2.3.b. Establishment of a single integrated multilateral comprehensive verifications system which *inter alia, ensures* that no nuclear weapons are produced.

2.3.c. Reduction of all conventional forces to minimum defensive levels.

2.3.d. Effective implementation of arrangement to preclude the emergence of a new arms race.

2.3.e. Universal adherence to the comprehensive global security system.

3.1 There has been a historically unprecedented militarization of international relations during the last four decades. This has not only enhanced the danger of nuclear war but also militated against the emergence of the structure of peace, progress and stability envisaged in the Charter of the United Nations.

3.2. To end this dangerous militarization of international relations, we must build a structure firmly based on non-violence. It is only in a non-violent democratic world that the sovereignty of nations and the dignity of the individual can be ensured. It is only in a non-violent world that the intellectual and spiritual potential of humankind can be fully realized.

3.3. The prospect of a world free from nuclear weapons should spur us to start building a structure of international security in keeping with the fundamental changes that are taking place in the world political, economic and security environment.

3.4. In a shrinking and interdependent world, such a structure has to be comprehensive, its components supportive of each other, and participation in it universal.

3.5. A world order crafted out of outmoded concepts of the balance of power, of dominance by power blocs, of spheres of influence, and of special rights and privileges for a select group of nations is an unacceptable anachronism. It is out of tune with the democratic temper of our age.

3.6. The new structure of international relations has to be based on scrupulous adherence to the principles of peaceful coexistence and the Charter of the United Nations. It is necessary to evolve stronger and more binding mechanisms for the settlement of disputes, regional and international. The diversity among nations must be recognized and respected. The right of each nation to choose its own socio-economic system must be assured.

3.7. Concomitant changes will be called for in the international economics order The interdependence of all the economies of the world makes for a symbiotic relationship between development in the South and stability and growth in the North. In a just and equitable order, access to technology and resources, on fair and reasonable terms will be assured. The gap between the rich and the poor nations will be bridged.

Index